Amaru Tincopa Gallegos

From the Voisin to the Mirage

100 years of French Aeronautical Presence in Peru

STRATUS

Published in Poland in 2016
by STRATUS s.c.
Po. Box 123,
27-600 Sandomierz 1,
Poland
e-mail: office@mmpbooks.biz
for
Mushroom Model Publications,
3 Gloucester Close,
Petersfield
Hampshire GU32 3AX, UK.
e-mail: rogerw@mmpbooks.biz

ISBN
978-83-61421-93-1

Editor in chief
Roger Wallsgrove

Editorial Team
Bartłomiej Belcarz
Artur Juszczak
James Kightly
Robert Pęczkowski

Colour Drawings
Teodor Liviu Moroşanu
JP Vieira

DTP
Stratus sp.j.

Printed by
Drukarnia Diecezjalna,
ul. Żeromskiego 4,
27-600 Sandomierz
www.wds.pl
marketing@wds.pl
PRINTED IN POLAND

Table of Contents

ACKNOWLEDGEMENTS

The author wishes to thank the following individuals and institutions for the invaluable support and collaboration in the making of this work:

Herbert Alcorta, Juan Arraez, Christian Barclay, Carlos Chávez Cateriano, Carlos Chinen, Cesar Cruz, Sergio De la Puente, José Fernández, Ricardo Iraola, Lewis Mejía, Danny Moeri, Santiago Rivas, Hilario Valladares, José Zlatar, Instituto de Estudios Histórico-Aeroespaciales del Perú (IEHAP), Dirección de Información de la Fuerza Aérea del Perú (DIRIN).

Photographic credits (in alphabetical order)
Herbert Alcorta, Juan Arraez, Carlos Chávez Cateriano, Cesar Cruz, Sergio De la Puente, Ricardo Iraola, Lewis Mejia, Danny Moeri, Santiago Rivas, José Zlatar, Instituto de Estudios Histórico-Aeroespaciales del Perú (IEHAP), Dirección de Información de la Fuerza Aérea del Perú(DIRIN).

INTRODUCTION

During the early years of the 20th century the Peruvian military was amidst a re-equipment and modernisation plan, which basically consisted in the complete rebuilding of the armed forces from zero after they were – for all practical purposes – completely destroyed during the so-called "Pacific War" fought between Chile and Peru from 1879 to 1883. Although this process emphasized ground and naval equipment, by the year 1911 a particular event – the flight of Peruvian aviator Juan Bielovucic over Lima, the capital of Peru, on January 15 – managed to impress the military figures, thus setting the first stone for what would later become one of the largest and most powerful air forces in Latin America. A few years later, the Peruvian Army and Navy both created their own aviation services, each of them following a major foreign country's experiences. Therefore, while the Navy relied on US experience with seaplanes for its Naval Aviation Corps, the Army decided to follow the French example, hiring a military aviation mission from that country, which already had a mission running the Army Officers' School in Lima. It was under this mission that the first dedicated military aircraft to serve in Peruvian military aviation arrived in the country. Even though this mission was short-lived, some of the aircraft it brought with it lasted for years at that city´s Military Aviation School, and their performance, ruggedness and ease of maintenance left a favourable impression on the (then) young officers that would later command military aviation.

When hostilities broke out between Colombia and Peru over the control of a large portion of the Amazon forest in late 1932, the subsequent arms race brought France once again as a major weapons supplier for the CAP, and a new influx of combat and training machines arrived on Peruvian soil. Particularly appreciated and remembered by their crews were the Hanriot and Morane types, which soldiered on for many years after their arrival, helping in the training of scores of Peruvian airmen, until their replacement by US-built aircraft by the mid-1940s.

Almost four decades later, new threats and strategic requirements of the *Fuerza Aérea del Perú* (Peruvian Air Force, FAP) called for the acquisition of a modern combat platform. Once again, the Peruvian Government selected a French product over several proposals from different countries, and it wasn't disappointed. The excellent safety record achieved by the Mirage 5P fleet, as well as the quality of the aircraft itself, opened the Latin American market for Avions Marcel Dassault, which was able to achieve several contracts for more than one hundred 'deltas' sold in South America alone. Almost two decades later, it was no surprise when in early 1980 the FAP became one of the first users of Dassault`s new product, the Mirage 2000 fighter, thus continuing with the legacy left by Salmsons, Moranes and Nieuports in past decades, and for many more years to come.

This is the story of the diverse French aircraft employed by the Peruvian military aviation from its beginnings to our days.

CHAPTER ONE

EARLY YEARS OF PERUVIAN AVIATION AND THE ARRIVAL OF THE FIRST FRENCH MADE AIRCRAFT

BIELOVUCIC AND THE VOISIN BORDEAUX: THE FIRST FLIGHT IN PERU

Juan Bielovucic Cavalie was an early aviator born in the city of Lima on June 30, 1889. A child product from the marriage between Croatian immigrant Juan Miguel Bielovucic and French lady Adriana Cavalie, Juan remained in Peruvian territory only until the age of eight, when a serious illness affected his father and forced the family to relocate to Dubrovnik, his father´s home town, on the Dalmatian coast (now Croatia). There was, unfortunately, nothing modern medicine could do for little Juan´s father and the head of the family passed away shortly after arriving in Europe. With the loss of the family provider the family was forced to relocate, once again, this time to the

A French postcard showing Bielovucic during his visit to Le Puy en Velay during July 15 and 16, 1911. (IEHAP)

A Young Juan Bielovucic seen aboard his Voisin moments before performing the first flight in Peru, in 1913. (IEHAP)

Pro-Aviation League members, military officers and curious bystanders gathered in the Santa Beatriz horse racetrack surround the odd looking machine shortly before Bielovucic's "unofficial" first flight. Lima, January 14, 1913. (IEHAP)

city of Paris, in France, where they could rely on the assistance provided by Mrs. Cavalie´s relatives in that city.

As young Juan grew, his admiration for the Brazilian aviator Antonio Santos Dumont – whom he knew personally – led him to follow his steps and soon he found himself enlisted in the "Voisin Brothers Aviation School" where he obtained his pilot license, the number 87, on June 10, 1910. The adventurous Juan Bielovucic was lured by the competitive spirit of the famous Air Meetings

Bielovucic flew for several days from the Santa Beatriz horse racetrack between January and March, teaching Peruvian students the basic concepts of flight. This picture shows a flight that took place on January 21, 1911. (Currarino)

held in Europe during those early years and, for the first time enrolled to participate in an air meeting. The Budapest Air Meeting, held in the Hungarian capital between June 5 and 15, 1910, was the first of these kind of events in which Bielovucic took part and was soon followed by others air meetings, circuits and raids such as the *Mourmelon-el Ferre* raid, held on August 15, 1910; the *Billancourt-Vannes-Malakoff* circuit held between August 28 and September 2, 1910 and, finally, the *Circuito Aéreo Internazionale*, held between September 24 and October 3, 1910, where Bielovucic was awarded with the fifth place in the "world distance classification".

On September 23, 1910, an attempt to be the first man to cross the Alps by using an airplane performed by Bielovucic's fellow countryman, aviator Jorge Chavez Dartnell, ended in tragedy when Chavez crashed aboard his Bleriot XI shortly before landing, sustaining serious injuries which lead to his demise four days later. A saddened Bielovucic vowed to complete his friend's ill-fated flight.

RETURN TO PERU

At the end of 1910, a group of Peruvian high-society figures – led by the Secretary of the *Ministerio de Guerra* (War Ministry, MG) General Pedro E. Muñiz – formed the *Liga Peruana Pro – Aviación* (Peruvian Pro-Aviation League), dedicating themselves to the development of aviation in the country. One of the first objectives of this league was to acquire machines for the future *Escuela Nacional de Aviación* (National Aviation School, ENA) and in order to do so they contacted Bielovucic with the aim of putting him in charge of the acquisition process of new aircraft, as well as to provide training for Army officers once he arrived in Lima. After signing the agreement, Bielovucic requested the assignment of the necessary funds to complete the agreement. When these were made available, Bielovucic embarked aboard the *SS Normandie*, along with two Voisin *type Bordeaux* aircraft and the French mechanics Henry Chailley and Pierre Schuck, both hired by Bielovucic to take care of the assembly and maintenance of the aircraft. At Panama, the party made a transfer to the SS *Ucayali*, and continued their journey towards Callao, where they arrived safely on January 8, 1911.

After the reception, Bielovucic began his duties, first of which was the search for an airfield to carry out their flying activities. Due to the absence of a proper airfield, it was decided to use the facilities of the city's old racehorse track, located in the Santa Beatriz district. There, the mechanics began assembling the aircraft with assistance from personnel from the "School of Arts and Occupations" of the Peruvian capital.

THE FLIGHT OVER SANTA BEATRIZ

With assembly completed Bielovucic was ready to commence with the test flight, which took place in front of a small number of LNA representatives at 1600 hours of January 14, 1911. The Voisin left the ground at and flew at an altitude of a couple of dozen meters for about twenty minutes, after which Bielovucic landed, albeit in a somewhat abrupt manner, damaging one of the landing gear wheels in the process, although this fact was barely noticed by the enthusiastic viewers.

After repairs the airplane sat ready for its "official" flight, scheduled for the next day. Therefore, at 1151 hours of January 15, 1911, Bielovucic took his aircraft to the air in front of a large crowd headed by Peruvian President Augusto B. Leguía as well as diverse high ranking officers and government figures, as well as numerous members of the Peruvian *socialite*, all gathered to witness this historic event. After completing this first "official" flight, Bielovucic performed several other exploits, one of them being the first flight over the Peruvian coast carrying a passenger, as well as a fly-past over the *Buques Armada Peruana* (Peruvian Navy Ships, BAP) cruisers *Almirante Grau* and *Coronel Bolognesi*, which were moored at El Callao bay. Moreover, the next day he took part in a fund-raising event, always from the Santa Beatriz's race track, with the aim of collecting funds for the acquisition of more airplanes.

On January 29, Bielovucic flew from Lima to the bay at Ancon, in what can be regarded as the first raid to the country "interior", which lasted about 40 minutes. During the following months, and always under the auspice of the LNA, Bielovucic began, along with French pilot and mechanic Henry Chailley, the training process of the first Peruvian aviators. Since the Voisin didn't have dual controls, Bielovucic had to seat in the small space behind the pilot's seat, now occupied by the student, and shout instructions to him. Such were the perils faced by early aviators during these years. The contract between Bielovucic and the *Liga Pro-Aviación* concluded in late March and, as it was not renewed, Bielovucic had to return to Europe to take care of his several commitments, on April 11, 1911. Mr. Chailley's contract, however, was extended for one additional month. Before his departure Bielovucic sold both Voisin biplanes to the Peruvian Government and, therefore, these became the first aircraft of French origin – or the first at all, for what matters – in the Peruvian service.

FRENCH AIRCRAFT ENTER PERUVIAN MILITARY SERVICE: CAPTAIN JUAN O'CONNOR'S BLERIOT FLIGHT

On the morning of December 2, 1915, *Ejército del Perú* (Peruvian Army, EP) Capt. Juan E. O'Connor took off aboard a Bleriot XI monoplane from a makeshift airfield set up in Bellavista, Callao, with the unprecedented mission – for the country´s military capabilities – to spot from the air the army drills scheduled to take place on that day. The small monoplane flew over the drill area before landing near the Army Officers School, in Chorrillos, where this officer issued a complete report of his flight. This was the first employment of an airplane with military purposes in Peruvian military history. The report issued by Capt. O'Connor to his superiors detailed his favourable impressions and considerations regarding aviation's future military use. As a direct consequence of this event the EP high command ordered a preliminary study for the creation of an army aviation force, which was completed by mid-1916.

Peruvian Army Capt. Juan Ernesto O'Connor poses next to his Bleriot XI shortly after completing one of the numerous observation missions flown on December 15 over the army drills held in San Juan de Miraflores, on the southern outskirts of Lima.

Taken from a newspaper of the era, this picture shows Capt. O'Connor's Bleriot during takeoff. (IEHAP)

Civilian and military figures give Capt. O'Connor a warm welcome as he jumps out from his Bleriot. (IEHAP)

The Farman III was used to renew Peruvian Army flight activities since the year 1915. Note the insignia from "El Palomar" flight school, previous owner of the aircraft, on the starboard wing. (Currarino)

Peruvian Army Lt. Guillermo Protzel del Castillo poses with the only aircraft available to Peruvian army aviators in 1919: a derelict Bleriot and the Farman III brought from Argentina by him.

THE FRENCH MILITARY AVIATION MISSION: THE MEN AND THE AIRCRAFT

After the creation of the *Servicio de Aviación del Ejército* (Army Aviation Service – SAE) in late 1918, continuous meetings were held between senior officers leading this entity and Government figures in order to reach a decision regarding which major foreign power would guide, train and equip this new branch of the Army. Thanks to the excellent results obtained by the French military mission, arrived in 1915, France got the balance tipped on its favor and, with the help and guidance of Lt. Guillermo Protzel del Castillo, as well as the civilians Manuel Sanchez

Ortega and Enrique Parodi Rodriguez, all three aviators trained in Argentina and the most respectable figures in the Peruvian aviation scene, the MG sent a proposal to Peruvian President José Pardo to give the French the responsibility to organize and provide material for the novel air service. This proposal found the favor of the government and shortly after a French aeronautical mission, to be attached to the one already present in the country, was set to arrive to the South American country.

By means of a renewable contract, for the amount of 3,750.00 francs[1] signed on May 1, 1919, the French Government committed itself to provide the Peruvian army with a select group of officers and technicians to organize, train and provide flight material for its aviation force. This Aeronautical Mission was composed by Colonel Paul Du Beaudiez[2], as chief of the Mission, Major Louis Couduret, appointed as Director of the *Escuela Militar de Aviación* (Military Aviation School, EMA), Lieutenants Alberto Chabrier, René Bremond, and José Romanet, assigned as flight instructors and, finally, Sergeant José Armant, as aviation mechanic. French General Vassal, head of the previously established French military mission, was appointed as a joint Director for both the Army and Aviation Missions. Along with this select group the first combat aircraft also arrived in the form of:

- 2 Morane-Saulnier MS.29C.1
- 1 Salmson 2A.2[3]
- 4 Caudron G.3
- 3 F.B.A. Schreck H
- 1 Farman F.60Torp Goliath
- 3 Georges-Levy GL.40HB.2.

Only ten machines from the afore mentioned list were actually purchased with funds issued by the Peruvian government with the remaining being acquired with money obtained from multiple donations done by notable Peruvian citizens residing in France.

After leaving Brest in late July the men and machines arrived at Callao aboard CVP[4] *Urubamba* on September 23, 1919. They received a warm reception, held in their honour both by the Peruvian authorities and the general population. In the days after their arrival the mission members proceeded to organize and prepare the beginning of their operations, starting by selecting a small field located in Bellavista as operations center. This decision, however, was soon to be regretted as that field was too small to provide a safe operations environment, becoming quickly overcrowded with aircraft as well as the related support equipment so it quickly became evident that a larger field would be required in the immediate future.

1 Enough funds to cover six first class tickets plus the mission members' expenses for one year, the extension of the contract.

2 A veteran from the Great War.

3 Later followed by a second machine.

4 *Compañía Peruana de Vapores*, Peruvian Steamship Co.

Maranga, December, 1919. A pair of Caudrons parked on the field await for a new sortie. (Currarino)

Bellavista, January 5, 1920. Lt. Gabriel Chabrier poses in front of one of the Caudrons brought by the French mission.

Maj. Coudoret and Lt. Romanet discuss upcoming operations in front of one of the makeshift Bellavista hangars. A Caudron G.3 sans cowling can be seen parked in the background. (Currarino)

Bellavista, January 5, 1920. Maj. Louis Couduret portrayed in the cockpit of one of the Caudron G.3 biplanes brought by the French Mission.

Lt. Louis Romanet, another member of the French mission, smiles to the cameraman from this Caudron G.3. Bellavista, January 5, 1920.

Bellavista, January 18, 1920. Salmson 2A.2 "Provincia de Pisco" and a pair of Caudrons seen in this field shortly before the tragic death of Du Beaudiez and Chabrier.

A view from the wrecked Caudron G.3 flown by Col. Du Beaudiez and Lt. Chabrier shortly after their tragic accident. (Currarino)

A Caudron in flight. (IEHAP)

With Salmson 2A.2 "Provincia de Pisco", purchased with the funds provided by the population from that Peruvian province, as a background Lt. Romanet poses next to the Mayor from the city of Pisco City and his wife.(Currarino)

The recently promoted Maj. Guillermo Protzel portrayed along some of his relatives in front of a Morane-Saulnier MS.35R from the Escuela de Aviacion Militar *based in Las Palmas in September 1923. (IEHAP)*

Unidentified officers and civilians pose in front of Morane MS.35R identified with military registry "131". Las Palmas, 1923. (De la Puente)

Lt. Roberto Bradley and student Jose Estremadoyro (handling the machine gun) aboard this F.B.A. Schreck H flying boat, identified with number "106", shortly before a training flight."Ancon", name of the navy seaplane station, appears painted in the aircraft nose. (IEHAP)

A Salmson 2A.2 takes off from a makeshift field in Arequipa, June 20, 1920.

Although of poor quality, this picture shows the elusive Morane-Saulnier MS.29C from EP aviation. Las Palmas, September 23, 1923. (IEHAP)

A clear picture of MS.29 number "123" seen at Las Palmas in September 1923. (DIRAF)

Interesting view of Salmson 2A.2 number "120" at Las Palmas. A Bristol F.2b type 15 and a de Havilland DH.9 could also be seen in the background. (IEHAP)

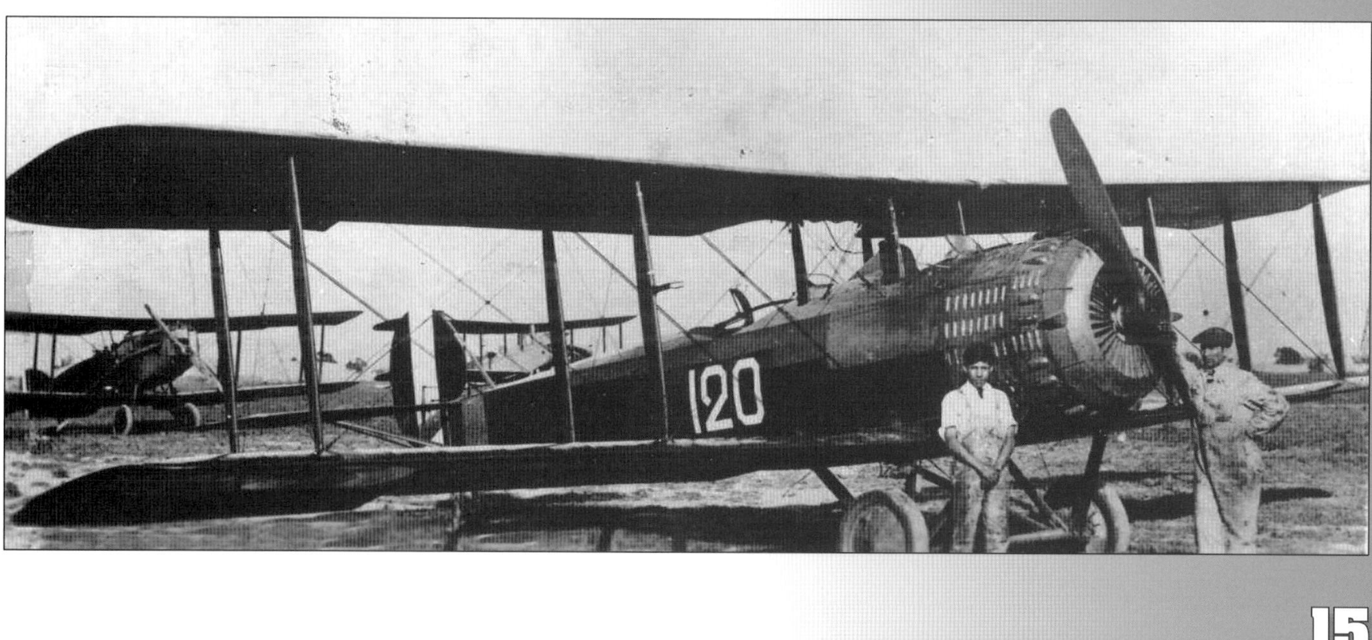

Las Palmas, 1923. A Morane-Saulnier MS.35C lifts off from the recently opened Escuela Central de Aviación Militar *"Jorge Chavez". Note the contrast between desert and foliage, typical Peruvian coast landscape. (De la Puente)*

Lt. Romanet (standing) and Sargeant Armand (in knees) appear in this rare view that shows the Salmson 2A.2 suplementary long range fuel tanks in advantage. (Currarino)

A rare image of Georges-Levy GL.40 flying boat seen shortly after completed its assembly stored in a half built hangar at Ancon seaplane base. (Currarino)

Another view of the rare Georges-Levy GL.40 flying boat in Peruvian service. This particular machine, identified with number "127", was employed by the chief of CAN, Cmdr. Juan Leguia Swayne, on its ill-fated raid to New York from February, 1921.

Three unidentified men aboard a F.B.A. Schreck F factory No. 888 from Naval Aviation Corps. Ancon, 1921. (IEHAP)

A DH. 9, an Avro 504R Gosport, a MS.35R, an Ansaldo A.300 as well as two "Dusters" appear in this picture taken in Las Palmas AB in February 1927. (Hagedorn)

THE DU BEAUDIEZ AND CHAUBRIER TRAGEDY

On January 18, 1920, and while looking forward to taking part in an air meeting scheduled as part of the celebrations for the country's new Constitution, the first accident in Peruvian aviation occurred, with tragic results. On that morning, Lt. Col. Du Beaudiez and Lt. Chabrier took off from Bellavista airfield aboard one of the Caudron G.3 from the School in order to take part, along other aircraft, in an air rally over the city. Looking to rendezvous with their already airborne comrades as quickly as possible, Beaudiez maneuvered his plane into a steep turn at only 50 m, causing the aircraft to spin and crash into the airfield. As a consequence of the serious injuries sustained during the impact, both aviators lost their lives before reaching hospital. This tragic event caused a great deal of grief amongst the population and their fellow comrades, as was evidenced by the large number of citizens who accompanied the remains of these unfortunate aviators during their burial, held with full military honors, at the "Presbítero Maestro" graveyard in Lima.

Col. Du Beaudiez was a Breton who had enjoyed a brilliant military career early in the Great War, distinguishing himself on the battlefield during the Madagascar, Senegal and Morocco campaigns which earned him the title of *Señor de la Legión de Honor*[5] in the year 1914. In September 1915 he was shot down over enemy lines and became a prisoner for the rest of the war.

BIRTH OF THE *CENTRO DE AVIACIÓN MILITAR DE MARANGA*

Given the limitations of the airfield at Bellavista, Peruvian authorities began the search for a new location for the mission's operations. After a short study, it was found that the "Fundo Maranga", a field located nearby, had all the required characteristics to serve as an airfield. Therefore, via *Resolución Suprema* (Supreme Resolution, RS) dated January 24, 1920, the state took over the 102,800 m^2 field and handed its administration over to the Mission under the name of *Centro de Aviación Militar de Maranga* (Maranga Military Aviation Center, CAM) and thus creating the country's first military aviation school.

Five hangars and all the basic structures required for operations such as offices, barracks, warehouses and a repair shop were quickly built over the field using the wood taken from the aircraft's empty containers. The field size and extension allowed for a 250 meters long by 30 m wide landing strip which, despite not being ideal, was enough for the students to perform their first flights.

On February 20, 1920, the official newspaper "El Peruano" published a RS on which the first list of students, both civilian and military, was announced, being composed as follows:

- Lt. (Infantry) Carlos A.Gilardi Vera
- Lt. (Artillery) José Lucas Ragúz
- Lt. (Infantry) Eduardo Bellantin Arenas
- Lt. (Infantry) Cesar A.Cossio Becerra
- Lt. (Engineers) Carlos Pro Arteaga
- 2nd Lt. (Engineers) Santiago Hard Campbell
- Midshipman Ismael Pro Arteaga
- Midshipman Augusto Belt Santisteban
- Mr. Alejandro Velasco Astete.
- Mr. Carlos Olavegoya Kruger
- Mr. Herbert Tweddle Valdeavellano
- Mr. Alejandro Gamboa Schuman
- Mr. Luis Garland Higginson

5 Legion Honour sir.

On June 12, 1920, shortly after the work on infrastructure was completed at Maranga, the Government officially authorized CAM operations, despite the fact that it had been active since March of that year.

THE MISSION'S FAILURE

The January 18 tragedy marked a breaking point in the relationship between the French Mission and Peruvian authorities, which quickly began to deteriorate. This situation was aggravated by the irresponsible attitude displayed by some of the French mission's members, which failed to live up to the expectations. As a matter of fact, during the rest of the year the French instructors did little or nothing to train the 13 students in their charge, due to the fact that they were busy carrying out "promotional flights" on behalf of the several French aircraft factories they represented as salesmen.

This situation proved unacceptable for the EP authorities, who were forced to send the most promising students overseas in order to complete their training, while civilian students were enrolled in recently created *Escuela de Aviación Civil at Bellavista*, managed by the Curtiss Aircraft Co. representatives. These measures were undertaken with the aim of avoiding a waste of human resources due to the negligence and indifference displayed by the French aviators. Meanwhile, the situation had also become of public knowledge and, as a consequence, the Peruvian Congress summoned the mission's head looking for explanations which, in the end, were considered unsatisfactory by the Peruvian authorities. This whole scandal resulted in several judicial processes, such as the interrogation of the *Secretario de Guerra*[6], General Antonio Castro, and his later demotion. Meanwhile, the Peruvian government exercised its right to rescind the contract, alleging a breach in the terms and – almost immediately – hired Royal Air Force (RAF) Captains Herbert Martin Sisson and Wilbur Bradley to take over the director and instructor roles, left vacant by the departing French.

Finally, it is worth mentioning that, due to bureaucratic delays, it was only in early 1921 when the *Cuerpo de Aviadores Navales* took delivery of the Farman, F.B.A. and George-Levy flying boats and floatplanes whose crates had been kept stored in the open at el Callao harbor since their arrival. With clearance from the *Ministerio de Guerra*, navy personnel proceeded to move the airframes from their storage location to Ancón where they were promptly assembled. Although there were some damages on the airframes from the exposure to moisture, these were repaired and the aircraft put into airworthy condition within days.

THE "NUNGESSER-COLI"

On December 19, 1927, French aviators Dieudonne Costes and Joseph Le Brix arrived at Las Palmas airfield in Lima aboard a Breguet Br. 19GR during their around-the-world flight started in France on October 14 that year. Their aircraft was christened as "Nungesser-Coli" in memory of French pilots who had become lost in the vastness of the Atlantic Ocean while trying to complete a non-stop transatlantic flight between Paris and New York aboard a Levasseur PL.8 christened "L'Oiseau Blanc". The aviators were escorted to Las Palmas by three aircraft from SAE which took off to welcome the French aviators. Upon landing Costes and Le Brix arrival was greeted by a large crowd led by high profile civilian and military authorities who had gathered at the airfield.

During their visit the French aviators performed a series of demonstrations for the Army and Navy officers present at Las Palmas to evaluate the aircraft but, despite the favorable impression left by the aircraft qualities and characteristics, no orders ensued. The epic flight by Costes and Le Brix lasted for a total of six months and four days and covered, up to its end on April 14, 1928, a total of 57,410 km.

---

6 War Secretary.

THE AIRCRAFT

Bleriot XI

Louis Bleriot was a French engineer, devoted to the production of a dozen aircraft between 1901 and 1909. In this year his company produced its most successful model to date, the Bleriot XI, designed by engineer Raymond Saulnier. Bleriot and this particular aircraft model became famous when they crossed the English Channel on July 25, 1909, covering the 40 kilometers between Calais and Dover.

The Bleriot XI became notorious in Peruvian aviation history due to the fact that it was the aircraft employed by Franco-Peruvian aviator Jorge Chavez Dartnell during his ill-fated Alpine crossing[7]. Another Bleriot, powered by the three cylinder Anzani engine, was used by Peruvian Carlos Tenaud during his flights over Lima on February 2, 1911. In December 1912, another Peruvian, Jose Ramón Montero, returned from France to his homeland bringing a Bleriot which he employed to carry out the first flight since the Bielovucic´s visit, a flight from Lima to the town of Pisco, which took place on February 12, 1913. Later, Montero donated this airplane to the Peruvian Army, which put it to good use during the Army exercises of 1915, when the first tactical flight took place performed by Capt. E. O'Connor.

On July 1913, the Italian aviator Miguel Rapini arrived in Lima to perform some propaganda flights aboard his Bleriot XI monoplane. Rapini flew for some time before his aircraft was damaged in an accident and, before returning to Italy, sold the damaged airframe to journalist Octavio Espinoza, who decided to return the aircraft to flying condition. The fragile monoplane was flown for a short period until the aircraft´s poor condition prevented any further use.

Henri Farman III

The Henri Farman III was a widely known aircraft from the early years of aviation. Designed and built by brothers Henri and Maurice Farman in 1910, the Farman III was a pusher-type biplane with a single forward elevator, powered by a 50 hp Gnome rotary engine. It was remarkably easy to fly and land, and was widely popular amongst young aviators.

This aircraft became tied to Peruvian aviation history in 1918, after when it was the first airplane flown by Peruvian Army officers Maj. Guillermo Protzel Del Castillo and Lt. Enrique Ruiz Espinoza after their arrival to "El Palomar" aviation school, in Buenos Aires – Argentina, where they arrived sent by the government to complete their flight instruction. Lt. Ruiz, unfortunately, was killed in an accident leaving Maj. Protzel as the only Peruvian student on Argentinean soil. After completing his training and before his return to its homeland, Maj. Protzel received a Farman III as a donation made by the Peruvian residents in Argentina to the newly born *Servicio de Aviación del Ejército*. After completing the formal arrangements the machine, baptized as *Teniente Ruiz* in honour of the deceased pilot, was brought to Lima by ship and, once reassembled, given good use by Protzel and other young SAE aviators. The operative life of the Farman was short as it was written off in an accident in late 1919.

Caudron G.3

The Caudron G.3 was a reconnaissance and light attack airplane developed by brothers René and Gaston Caudron from the earlier G.2. It could be fitted with one light machine gun as well as light hand-held bombs. This particular model saw service during the first stages of the Great War,

7 On September 23, 1910, Chavez took off from Ried-Brig, in Switzerland, and successfully made his way through the Simplon Pass. While approaching to land at Domodossola, Italy, his Bleriot XI suffered a structural failure and plunged to earth, and the unfortunate aviator suffered severe internal injuries which caused his demise a few days after.

flying mostly reconnaissance missions where its strong construction was well appreciated at the frontline.

Four Caudron G.3 dual-control biplanes were amongst the machines brought by the French as part of their Military Aviation Mission. The service record of these machines was less than stellar, with one lost along with its two crewmen at Bellavista airfield on January 18, 1920 and the three others grounded shortly after the departure of the French mission, as Peruvian flyers refused to fly the type due to its rather dangerous flying characteristics.

Salmson 2A.2

The Salmson 2A.2 was developed in late 1916 following a requirement by *Armée de l'Air* which was looking for an indigenous multi seat reconnaissance and light attack aircraft which could replace the Sopwith 1½ Strutter, at that time the backbone of the tactical reconnaissance arm. As Societé des Moteurs Salmson had previously produced the British biplane under license the resulting aircraft, identified as Salmson 2A.2, was heavily influenced by the Strutter and, after its entrance in service in late 1917, became one of the most successful reconnaissance aircraft of the Great War.

The Salmson 2A.2 was among the several types brought to Peru by the French Military Aviation Mission. While originally a single machine arrived with the French delegation in Lima during 1919, a second airframe did so in the autumn of 1920 purchased with the funds provided by the population of the Pisco province, located south of Lima, and accordingly christened as "Provincia de Pisco". Meanwhile the original Salmson flew with no particular markings until the year 1923, when the number "120" was applied on both sides of its fuselage. This particular machine was fitted with a T.S.H.[8] telegraph system which enabled the first communications between an aerial moving vehicle and a fixed position in the country.

The Salmson 2A.2 enjoyed a good reputation amongst Peruvian flyers thanks to its rugged construction and pleasant handling characteristics. If there was any criticism against it, however, it was regarding the somewhat excessive distance between the crew positions, which made communications between pilot and navigator almost impossible during flight. Both Salmsons served until late 1925 when the lack of spare parts and the excessive wear and tear of their structures made any further use impossible.

Morane-Saulnier MS.29C.1

The MS.29C.1 fighter was a parasol-type monoplane design that could be powered by any of three different engine types: Gnome *Monosoupape* rotary nine cylinder engine of 150 hp, 120 hp Le Rhone 9J rotary engine or a 130 hp Clerget radial engine. Pressed into service with the AdA in late 1917, this type evidenced structural failures which were cause of some accidents which led it to be relegated to advanced training duties shortly after its debut.

A pair of Morane-Saulnier MS.29s were amongst the plethora of aircraft brought by the French mission in 1919, becoming the first fighters to serve with Peruvian military aviation. A hot mount, the MS.29 was only flown by the more experienced pilots from the SAE. During the inauguration ceremony of the *Escuela de Aviación Militar* in late 1923, one MS.29 was flown by 2nd Lt. Alejandro Velasco Astete who offered a great display for the public despite ending with the aircraft sitting on its nose after landing, fortunately without serious damage. By 1925 the lack of spares and accidents had grounded both fighters, which were soon scrapped.

---

8 Telegraph Sans Hil, wireless telegraph.

Morane-Saulnier MS.35ET.2

Widely known as the Morane "AR" during the First World War, the Morane-Saulnier MS.35, as it was known after the armistice, was a two-seat, parasol-type monoplane of conventional design, with the crew seating in tandem in an open cockpit. This monoplane was of wooden construction covered with aluminum at the front of the fuselage and fabric over the rear. It had three major variants, each one differentiated by the type of engine employed: the MS.35R employed an 80 hp Le Rhone 9c radial engine, the – C a Clerget 9c, and the – A an Anzani. The aircraft's good flying characteristics made it very popular amongst pilots and it was purchased by many air forces in the 1920s.

Four Morane-Saulnier MS.35Cs were purchased by the Peruvian Government through the French mission in late 1920. These arrived in mid-1921, becoming the first real dedicated trainers to be employed by Peruvian military aviation. After the dismissal of the French mission, they were moved from Maranga to Las Palmas airfield, where they enjoyed a rather long service life, flying for many years and training dozens of new pilots until late 1928, when the last airworthy machine was finally grounded.

F.B.A. Schreck H

Franco British Aviation (F.B.A.) developed the Schreck model H hydroplane during the Great War. This model was extensively used by both French and Belgian air forces during the conflict and was exported to numerous countries after the war.

The pair of F.B.A. flying boats brought by the French mission suffered the perils of bureaucratic mess, characteristic of the first years of military aviation in Peru, being kept stored in their crates until the authorities took a final decision regarding their future operators. By the time the F.B.A. became finally assigned to the CAN and assembled in early 1921, damage caused by almost a year of exposure to the humid conditions found in el Callao was discovered in their structures and, consequently, the airplanes saw only brief service with the CAN's *Escuela de Hidroaviación*[9]. Assigned with the numbers "121" and "122", the F.B.A.'s operated in the training role alongside the Norman-Thompson NT.2B and Curtiss MF "Seagull" from the school until the summer of 1922 when lack of proper maintenance as well as the tear and wear on their structures forced their retirement.

Farman F.60 Torp Goliath

Developed by Henri Farman in 1919, the F.60 was an enormous bomber/transport for its time. A floatplane torpedo-launching version was soon developed after the war, and offered to various countries.

A single Farman F.60 Torp was brought to Peru by the French mission in 1920 but remained unassembled until 1921 when it was finally assigned to the *Cuerpo de Aviadores Navales*. Even then the aircraft saw limited to null service as it was seldom flown from Ancón, mostly due to the lack of qualified crews, and was eventually withdrawn from service and scrapped in late 1922.

George-Levy GL.40HB.2

An elegant 3 seat flying boat, the Georges-Levy GL.40HB.2 (*Hydravion de Bombardment* or Bombardment Seaplane) was designed by Maurice Jules-Marie Le Pen and Ing. Blanchard and saw service with the French, Belgian, Portuguese and Finnish air arms, with mixed results. While the French regarded the GL.40HB.2 as a successful machine, thanks to the type's good service record, the Finns had a different opinion after losing three out of a dozen aircraft, causing a number of fatalities.

9 Seaplane school.

Three machines arrived to Peru in late 1919 but, like the other seaplanes and flying boats arrived with the French mission, they remained unassembled for a long period of time, and as a consequence some of their components were already damaged by the time they were finally assigned to CAN by the end of 1920, and were ferried to Ancón where it was decided to assemble only two out of the three airframes, which were assigned with numbers "127" and "128", while the third was kept as source for spare parts. On February, 1921, Cmdr. Juan Leguía Swayne planned an ambitious Ancon–New York flight, employing one of the George-Levy flying boats due to their appropriate range. The flight was planned to follow the route Ancon-Santa Elena (Ecuador) – Buenaventura (Colombia) – Panama City (Panama) – Colon (Panama) – Kingston (Jamaica) – La Habana (Cuba) – Key West (Florida, USA) and finally New York. After planning was completed, Leguía left Ancon aboard George-Levy number "127" at noon on February 17, accompanied by Walter A. Jawielski as mechanic. However, after four hours flight a radiator malfunction forced the pilot to make a stop in Salaverry harbor, Trujillo. Close examination revealed damage to the radiator and, consequently, a replacement had to be ordered to Ancon, which was delivered after five days. However, as one problem was solved, another one appeared. Shortly after take-off a severe vibration forced the pilot to land, only to discover that his propeller was damaged. Another five days passed until a replacement propeller was delivered and fitted to the plane, and only then Leguía and Jawielski were able to carry on with their journey heading towards Talara where the streak of bad luck continued. A wave struck the flying boat's left wing during landing and caused severe damage that took days to repair. Only on March 15 were the airmen able to continue north reaching Santa Elena, in Ecuador, a few hours later. However, in this location the Peruvian aviators found unexpected hostility from the local authorities who banned the aircraft from entering harbor and even rejecting their request for refueling the aircraft in order to continue their journey. Frustrated and unable to proceed Cmdr. Leguía took the most reasonable course of action and loaded its flying boat aboard the SS *Europa* steamer, which was heading towards el Callao, and returned home.

Both flying boats were kept in CAN strength until the summer of 1923 when lack of spares and the continuous damage to their wooden structures caused by the excessive humidity of the Peruvian coast forced them to be withdrawn from service and scrapped.

Las Palmas, December 19, 1927. A series of pictures showing the arrival of the Breguet BR.19GR baptized "Nungesser-Coli", flown by French aviators Dieudonne Costes and Joseph LeBrix.

Las Palmas, December 19, 1927. A series of pictures showing the arrival of the Breguet BR.19GR baptized "Nungesser-Coli", flown by French aviators Dieudonne Costes and Joseph LeBrix.

CHAPTER TWO

THE INTER WAR YEARS: Hanriots, Moranes, Nieuports & Potez

As a part of the arms race started between Colombia and Peru in the context of the so-called "Putumayo campaign", the Peruvian Government sent purchase commissions overseas looking to close the gap in numerical advantage gained by Colombia after this country´s large order for amphibious fighters and bombers[1]. One of such commissions arrived in France where it visited the Potez, Nieuport, Morane-Saulnier and Hanriot factories, amongst others, to evaluate aircraft offered by these firms. This translated in the signature, from March 5 to the 8, of several contracts between Peruvian government and the *Societé Generale de Aeronautique*, represented by Mr. Paul Dietrich, for the delivery of 37 aircraft, divided between fighter, light attack and trainers whose, added to a similar number of machines acquired in England and Italy, would enable the *Cuerpo de Aviación del Perú* (Peruvian Aviation Corps, CAP) to match the acquisitions made by the Colombian military aviation.

The cease fire signed on May 24 of 1933 between Colombia and Peru transformed into a permanent peace agreement a year later, causing a normalization of relations between the two countries and the military demobilization in the area. This meant that there was no longer a need for many of the aircraft purchased to be used in anger in the Amazon forest but nevertheless, the fresh influx of equipment, particularly the trainers, enabled the CAP to vastly increase its strength and capabilities during the years to come.

Hanriot H.438

This multipurpose biplane was an export development from the Hanriot H.433 but powered by a more powerful 330 hp Lorraine Mizar 7Me seven cylinder radial engine replacing the original 230 hp Mizar 7Mc. Its metal construction, composed of a steel structure covered with aluminum panels, made it a rugged machine, while its sturdy landing gear allowed its use from unprepared fields. The CAP purchased six H.438 aircraft[2] in early April 1934 at a unit price of 270,000 francs. After arrival they were assigned to the *V Escuadrón de Instrucción* (V Instruction Squadron, V IE) based at Las Palmas AB in Lima and placed under the command of *Escuela Central de Aviación* (Central Aviation School, ECA), where they served as intermediate trainers.

November 1937 was a particularly bad month for the CAP Hanriot fleet. On the 13th a H.438 identified with the CAP number E/IA-3A suffered an accident at Fundo "La Chacarita" when 2nd Lt. Alejandro Gamon crashed while flying at low level. Then, on the 24th, Hanriot number E/IA-2A was heavily damaged when its pilot, 2nd Lt. Carlos

The future Peruvian military aviation hero, José Abelardo Quiñones Gonzales, pose as a cadet sitting on the landing gear of a Hanriot from Escuela Militar de Aviación *Las Palmas 1937. (IEHAP)*

1 During late 1932 and early 1933 Colombian *Aviación Militar* ordered a total of 27 Curtiss Cyclone Falcon and 24 Curtiss Model 35A Hawk II.

2 Identified as follows: Construction/Number 3 (French civil registry F-AJTJ number 2534), Co/Nu 4 F-AJTK No 2535, Co/Nu 5 F-AJVX No 2584, Co/Nu 6 F-AJVY No 2585, Co/Nu 7 F-AJVZ No 2586, Co/Nu 8 F-AJXM No 2587

Fine ¾ view of a Peruvian Hanriot H.438 at Hanriot factory at Bourges. Las Palmas, February 1934. (Ledet)

Lineup of Hanriot H.438 at Las Palmas shortly after delivery. At least four Morane-Saulnier MS.316 can be seen in the background. (IEHAP).

Several training aircraft from Escuela Central de Aviación *can be seen in this picture taken at Las Palmas in the summer of 1936. Among the different aircraft at least two Hanriot H.438 and a single MS.231 can be seen behind this Travelair E-4000B. (Gentili)*

A formation of CAP Hanriot H.438. (Goto)

Lt. Ciriani and an unidentified sub officer are pictured aboard a Hanriot H.438 from V EI. The Squadron mascot can also be seen seated on top of the wing. (IEHAP)

A Hanriot H.438, two Curtiss O-1B Falcon and one Morane-Saulnier MS.316 prepare to take off during training operations at Las Palmas in 1936. (IEHAP)

Insúa, misjudged its landing angle. On both occasions the crews walked away with only bruises.

During 1939 and 1940, the three remaining aircraft were sent to the *Fábrica Nacional de Aviones Caproni* in Las Palmas to be inspected. It was determined there that only two of them were worth repairing as the third, identified with number E/IA-6A, was in poor condition and was officially withdrawn from service on April 27, 1940. After overhaul, the two remaining aircraft received new numbers, I-2-7 and I-2-8, and returned to active service with the 5to *Escuadrón de Instrucción* (5th Instruction Squadron – 5 EI), serving in that unit until 1942 when both airframes were decommissioned and scrapped

A post card picturing the formation flight by the students from the "Almirante Miguel Grau" class in 1936. A total of six Hanriot H.438 can be seen closing the formation. (Currarino)

June 7, 1936. Aircraft from Escuela Central de Aviacion perform a flyover over Bolognesi Plaza, in Lima, during celebrations for the national Flag day. From left to right, three H.438, four MS.316, two Curtiss N2C Fledgling, one Vought O2U-1E and two H.438. (Currarino)

Morane-Saulnier MS.231ET.2

The search for an advanced trainer led the Peruvian technical commission to visit the Morane-Saulnier factory in Puteaux in early 1933 to evaluate the MS.230 parasol-type monoplane, widely employed in its many variants by the *AdA* as well as other air forces. Favorably impressed by its performance and sturdy construction, the commission issued a report recommending the purchase of a single machine to be assigned to the advanced training section from V EI. It is worth notice that, to simplify logistics and maintenance, the technical commission requested the French firm to modify the aircraft in order to be fitted with the same Lorraine Mizar 7b engine powering the Hanriot H.438, simultaneously ordered. Morane-Saulnier proceeded with the conversion with the resulting sub-variant receiving the factory designation of MS.231ET.2.

MS.231 civil registry F-AJUK seen parked at the Morane-Saulnier factory airfield in Puteaux shortly before its delivery to Peru. (Currarino)

3rd year CAP cadets pose with the single CAP´s MS.321 christened as "Azul". A young Jose Abelardo Quiñones Gonzales rests in the port landing gear strut. (IEHAP)

Students and instructor stand in front of the "Azul", the single Morane-Saulnier MS.213ET.2 from the V EI. (IEHAP)

A smiling 3rd year cadet Rafael León de la Fuente poses in front of Morane-Saulnier MS.231 identified with CAP registry E/IA-1B. Note Morane-Saulnier factory logo in the engine cover. Las Palmas, 1937.

The aircraft was identified with factory construction number No.3 and received French civil registry F-AJUK and number 2897 before its delivery to the CAP in the summer of 1934, where it was pressed in service as E/IA-1B with the 4th section (fighter and bomber training) from the V EI. During its service with the CAP the *Azul* (blue), as it was popularly known thanks to its overall finish color, trained dozens of fighter pilots until 1939 when it was written off in an accident and withdrawn from service.

Morane-Saulnier MS.316ET.2

The Morane-Saulnier MS.316 was the export denomination given to the Morane-Saulnier MS.301 twin seat basic trainer parasol monoplane, powered by a 100 hp Lorraine 5Pa five cylinder radial engine. Six of these machines were ordered by the Peruvian Government in March 1933, with the deliveries completed by the factory by late June of that very same year. The airframes, identified with correlative factory numbers N° 1 to N° 6, were assigned to the 1st section (primary training) from V EI with correlative CAP numbers E/1P-1A to E/IP-6A. Due to their silver dope finish these aircraft were affectively known as the *Plateados* (silvers) by pilots and ground personnel.

While pilots found this aircraft's handling pleasant and trouble free, there were still accidents. On May 13, 1935 2nd Lt. Enrique Escurra was killed aboard Morane-Saulnier MS.316ET.2 number E/IP-6A when he lost control of the aircraft during a low level maneuver and crashed on a field near Las Palmas AB. Another Plateado was lost along with its crew a year later when, on June 19, 1936, 2nd Lt. Patron and Officer Cadet Manuel Morales were both killed when their aircraft entered a spin and crashed in Villa auxiliary airfield, located a few miles south west from Las Palmas.

In October 1938 the four surviving Moranes were sent to the Caproni Aircraft Factory in Las Palmas for a complete overhaul, which included a change in the power plant, replacing the original 100 hp Lorraine 5Pa engine with a 125 hp Kinner B-5. Logistics was the main reason behind the selection of this engine as it was the chosen powerplant to be fitted on the Caproni Ca.100PR basic trainers' series being built by the FNA at that time. As consequence of the increased output of the Kinner engine, a slight increase of performance was recorded on the first flights.

During September 1939 the Caproni factory began the progressive delivery of the newly overhauled machines which were assigned with new ECA numbers according to their arrival, becoming identified as I-1-1, I-1-2, I-1-4 and I-1-16. Accidents, however, took a toll on their number and, on August 19, 1940, I-1-4 was written off after its pilot misjudged the landing approach and ended crashing against a fence at the end of Las Palmas AB runway. At the beginning of 1940 and with the arrival of enough Caproni Ca.100PR basic trainers the *Plateados* were reassigned into the 2nd Section (intermediate training) from *V EI* and re-numbered as I-2-1, I-2-3 and I-2-13. In November 1941, however, I-2-3 was written off and its pilot seriously injured after it stalled during takeoff.

The arrival of US built trainers like the PT-17 Kaydet, supplied under the auspice of the Lend-Lease program in late 1942, put an end to the career of the MS.316 trainers with the CAP. The rugged Moranes, however, were sent to the *Arsenal de Aeronáutica* where their structures and engines were inspected and repaired, and then were donated to the *Escuela de Aviación Civil* (Civil Aviation School) at Collique airfield on the city outskirts. There, the Moranes – now in civilian markings – flew until mid 1947 when the wear and tear of the last surviving machine, identified with civil registry OB-GGJ, forbid any further use.

The first of the six Morane-Saulnier MS.316ET.2 ordered by the Cuerpo de Aviación del Perú *seen parked at Puteaux shortly after its completion. The aircraft sports both the* Arriba Siempre Arriba *as well as the V* Escuadrón de Aviación *unit acronyms. (Liebert)*

Three-quarter view of the view from the same aircraft. (Liebert)

September 23, 1936. Five Morane-Saulnier MS.316 take off in chevron formation from the grass runway at Las Palmas airfield during the ceremony for the Peruvian aviation day. (Currarino)

Cadet Alfredo Salazar Southwell in the instruction seat of this Morane-Saulnier MS.316 in 1935. (Currarino)

Instructor and students from ECA "Jorge Chavez" pose in front one of the Morane-Saulnier MS.316 from the school in 1936.

Young cadets from the first year of studies pictured next to their basic flight instructor with an Morane-Saulnier MS.316 as background.

Members from the "Coronel Francisco Bolognesi" student Class, mascot included, portrayed in front of a Morane-Saulnier MS.316ET.2 from the V EI in late 1935. (IEHAP)

A Morane-Saulnier MS.316ET.2 number E/1P-3A and "ASA" painted on its fuselage appears next to a Douglas O-38P. "ASA" stands for "Arriba Siempre Arriba" ("Up, Up to the stars"), last words from Peruvian aviation icon Jorge Chávez Dartnell. (De la Puente)

Caproni Peruana Aircraft Factory, Las Palmas 1938. A lone Morane-Saulnier MS.316 stands inside the factory hangar shortly after it was fitted with the Kinner B-5 five cylinder 125 hp engine. (Caproni via Apostolo)

Above: *A pair of unidentified civilians stand in front of a pair of Morane-Saulnier MS.316 during an open doors at Las Palmas AB in early 1941. Note the different shape of the nose of the Kinner engine version. (Velasquez)*

Left: *2nd Lt. Oscar Rizo Patron seen portrayed in front of one of the Morane-Saulnier MS.316 trainers from the V EI. This young pilot lost his life in one of these aircraft on June 19, 1936. (IEHAP)*

Instructors, students and technical personnel portrayed in front of a Kinner engine MS.316ET.2 in Las Palmas AB during late 1939. Note the aircraft's non standard engine cover. (IEHAP)

Morane-Saulnier MS.316ET.2 CAP number I-1-2 began its takeoff run from Las Palmas grass runway on Autumm 1939. (De la Puente)

Accidents took a toll on the ageing Morane fleet. This Morane-Saulnier MS.316ET.2 identified with number I-1-4 ran against a fence outside the end perimeter of Las Palmas AB due to a pilot's mistake in December 1939. (IEHAP)

Left and below: *With the arrival of the Caproni Ca.100PR primary trainers, the remaining MS.316ET.2 were replaced in the primary training and assigned to intermediate training duties.*

Morane-Saulnier MS.316 number I-2-3 stalled after takeoff and sustained heavy damage. Miraculously, its only occupant walked away with only small bruises.

After being withdrawn from service with the CAP the surviving MS.316s were assigned to the Civil Aviation School. This Morane-Saulnier MS.316ET.2, identified with civil registry OB-GGJ, flew until 1947. (Currarino)

Potez 390A.2 and 391A.2bis

The Potez 390A.2 was an export variant derived from the Potez 39 two seat reconnaissance and light attack aircraft developed in 1933 to fit the CAP technical commission requirements for a full metal construction reconnaissance and light attack aircraft, powered by a radial or in-line engine on the 500 hp range and capable to operate either with conventional landing gear or with floats with a minimum loss of performance.

In order to compensate for the expected drop of performance on the floats-equipped version the company replaced the original 500 hp Hispano Suiza 12Hb engine with the more powerful 550 hp Lorraine-Petrel 12Hars, a change that gave the aircraft a whole new nose appearance. Unfortunately the trials of the float-equipped version, which took place at the Potez-CAMS factory in Sartrouville, showed poor engine performance as well as insufficient directional stability and as consequence the CAP officers in charge of the evaluation requested Potez to make modifications to the aircraft power plant and vertical surfaces in order to meet the requirements established by the technical commission. Consequently, Potez offered to modify the aircraft and in autumn of 1933 the commission arrived once again at Sartrouville to observe the tests of the updated model,

Two views from Potez 390A.2 fitted with floats during tests at the Potez plant in Sartrouville, summer 1934. (Ledet)

identified as Potez 391A.2bis. This upgraded variant was externally almost identical to the original model 390A.2, except for the fitting of a larger tail surface, a measure taken to improve directional stability of the aircraft which was hampered when it was fitted with the large floats. It was not the only difference, as the engine cowling housed a more powerful 720 hp Lorraine-Dietrich 12Hdrs 12 cylinder in-line engine. While these changes notably improved performance and handling, budget restrictions forced the CAP to abandon this upgrade to the whole order and therefore only three Potez 390s were upgraded to this version while the remaining were delivered with the less powerful engine and had the regular tail configuration.

Meanwhile in Lima the purchase of both the Potez as well as other French aircraft suddenly came under attack from a number of senior CAP officers whom argued, with some valid reason, that some of the aircraft being acquired were only prototypes and not in actual service with the French *Armee de l´Air*, a situation considered to be unacceptable to them specially due to the high prices being paid for such material. The deal went ahead, nevertheless, and CAP representatives took delivery of the first airframes at the Potez factory at Sartrouville in early March 1934. A month later the first two Potez observation and light attack monoplanes arrived in Callao, with the rest of the order delivered by the factory at a rate of two aircraft per month. After arrival the Potez were assigned to the newly formed *4to Escuadrón Mixto* (4th Mixed Squadron – 4 EM), based in "Alferez Carlos Huguet" seaplane base in Ancón, where they formed a seaplane fighter and attack unit alongside the Nieuport-Delage NiD.121C.1 fighters, starting an intensive training cycle in preparation for the eventual deployment to Iquitos, in the Amazon region. The peace between Colombia and Peru, however, was signed shortly after the delivery of the first Potez and, therefore, in order to facilitate the peace process between the two countries the CAP High Command discarded the plans aimed to the reinforcement of the aerial units in the Amazon region, disbanding the 4 EM and ordering its aircraft to be sent to Las Palmas AB.

In service

Back in Las Palmas the Potez became assigned to a new unit, the newly organized *Primer Escuadrón de Bombardeo y Observación* (1st Bombardment and Observation Squadron – 1 EBO). Composed by three escadrilles, with three Potez each, plus a reserve aircraft, the 1 EBO was located under the II EA command and its aircraft were identified with the correlative numbers II/1-BO-1B to II/1-BO-9B. The remaining pair of aircraft was assigned to the ECA to serve as trainers for new crews.

On April 25, 1936, a pair of Potez from 1 EBO were flown to the Peruvian city of Arequipa, located in the Andes mountain range at about 1000 km south east distance from the capital Lima, in order to participate in the ceremony for the city´s new airport inauguration.

In June 1936 the *Comandancia General de Aeronáutica* (High Command – CGA) introduced a number of organizational changes to the air force beginning with the change in the denomination of the air force from *Cuerpo de Aviación* to *Cuerpo Aeronáutico del Perú*. Also, amongst the numerous changes introduced as part of the process was the creation of four "Aerial Regions", distributed along the territory, each one assigned with a corresponding *Grupo Aéreo* (Air Group – AG). Accordingly, the First Aerial Region covered the Peruvian north and north-west borders; the Second Aerial Region covered the central and southern sectors of the country as well and its borders while the Third Aerial Region was tasked with the protection of the pacific coastline as well as the territorial waters. Finally, the Fourth Aerial Region was tasked with the defense of the east and north-east border, in the Amazon forest.

First Aerial Region headquarters was located in "Teniente Coronel Pedro Ruiz Gallo" AB in the city of Chiclayo. Located about 720 km north of Lima, this recently opened base was the home of *Primer Escuadrón de Aviación* (First Aviation Squadron, 1 EA) which, following the newly established territorial organization doctrine, was composed by a bomber, a fighter and a reconnaissance squadron. In order to fulfill the reconnaissance and light attack duties of this new unit, the 1 EBO was reorganized around five machines and put under the 1st EA command. On October 5, 1936, the Potez left Las Palmas heading for "Teniente Coronel Ruiz" AB, along with nine Caproni Ca.111 and nine Caproni Ca.114 fighters, also part from the 1st EA aircraft complement. Meanwhile, the five remaining Potez were held as reserve in Lima.

As the Peruvian Air Force Strategic Plan issued in January 1937 required also the deployment of a reconnaissance unit to be put under the command from 2nd Aerial Region in order to provide strategic reconnaissance for the naval and ground forces on the south of the country, the remaining Potez were allocated under this command and five aircraft from 1 EBO flew from Las Palmas AB to "Capitán Guillermo Protzel del Castillo" AB in Vitor, department of Arequipa. As a part of their training, both detachments from this unit carried out flights to different airfields inside the country during which the Potez demonstrated its good climbing capabilities and adequate range.

A tragic event took the life of a very promising CAP officer on September 14, 1937, when a Potez 390A.2 number II/1-BO-2 flown by 2nd Lt. Alfredo Salazar Southwell as part of a five ship formation doing rehearsals for an upcoming air meeting to be held a few days later, suffered an engine fire which forced its pilot to abandon the formation and, after giving orders to his observer/gunner to abandon the aircraft, head to the nearby coast in an attempt to avoid the civilian residential areas below. Unfortunately for Southwell his Potez was soon engulfed by the flames and the pilot crashed to his death on an empty Army barrack in the Miraflores district, fortunately without causing civilian property or lives losses.

The end of the road

The chaotic situation that affected the French aircraft industry during the second half of the thirties had repercussions on the operability levels of the users of French aircraft abroad, and the CAP was no exception. Limited availability of spare parts, especially engines, mostly caused by disturbances in the production at the French factories due to labour conflicts, but also due to the

Sartrouville, summer 1934. Potez 390A.2 on the Seine river during trials. (Ledet)

The larger vertical fin area identifies this machine as the somewhat improved Potez 391A.2Bis, photographed during trials by the CAP technical commission at Seine River near Sartrouville. Note Peruvian markings already in place. (Arraez)

A diverse array of CAP aircraft parked at Las Palmas AB in late 1934. Three Potez 391A.2 and at least one NiD.121C.1 can be seen among several Fairey Fox, Curtiss Falcon and Vought Corsairs, certainly an interesting lineup. (Davis)

Nine Potez purchased by CAP at Las Palmas, March 1934. (Ledet)

The small air scoops on the engine cowling identify 4-O-66 as of the three Potez 391A.2Bis delivered to the CAP. This Potez wears the numbers assigned to the 4[to] Escuadron Mixto *aircraft. (Liebert)*

Profile view from of the same aircraft. (Liebert)

A Potez 390A.2 from the 4to Escuadrón Mixto. *(IEHAP)*

Montage showing a formation of CAP Potez 390A.2.

Airmen from 1 EBO pose for the camera in front of one of the Potez 390A.2 from the unit in early 1936. Of interest is the large size of the aircraft as well as the Caproni Ca.111 bombers parked in the background. (IEHAP)

Officers from the "Coronel Francisco Bolognesi" class pose along one of the ECA Potez 390A.2. Noteworthy are the under-wing bomb racks as the large size of the aircraft.

The whole air and ground crews from Primer Escuadron de Aviacion *pictured in front of two Potez and three Caproni Ca.114 fighters at "Teniente Coronel Ruiz" airbase in Chiclayo, November 1936. (Currarino)*

Officers and sub officers from Primer Escuadron de Aviacion *pose with one of the Potez 390A.2 from the unit in January 1937. (Currarino)*

A magnificent view of a pair of Potez 390A.2 overflying the Argentinean warships ARA Rivadavia *and ARA* 25 de Mayo *moored at Callao bay. La Punta peninsula can also be noted in the background. Lima, January 1937. (Goto)*

2nd Lt. Alfredo Salazar Southwell and S.O. Carlos Fajardo stand next to one Potez from ECA at Las Palmas in September 2, 1937. A few days later Southwell was killed aboard one of such aircraft. (IEHAP)

Lineup of Potez from both 1 Escuadrilla de Bombardeo y Observacion *as well as from ECA. Las Palmas, 1937. (Goto)*

Chiclayo, late 1941. President Manuel Prado Ugarteche (in suit and hat) walks in front of a Potez while inspecting "Teniente Coronel Ruiz" airbase along the chiefs from the Peruvian Army, Aviation, Navy as well as the US Military mission. (Currarino)

CAP mechanics and technical officers use this Potez 390 as instructional airframe, after it was withdrawn from active duty. The data tag in the tail suggests that the Peruvian aircraft were mere conversions from Potez 39 from the production line. (IEHAP)

rearming requirements from the French AdA, caused a negative effect on the aircraft from 1 EBO, with the situation becoming critical in late 1938 when lack of parts had grounded nearly the total of the unit aircraft strength. In Vitor, the remaining aircraft were quickly disposed and scuttled, disappearing without leaving trace and, while nearly all the aircraft based in "Teniente Coronel Ruiz" AB suffered the same fate, by cannibalization of other aircraft ground crews managed to keep a single Potez in airworthy condition to be employed as a target tug by the fighter units with base in Chiclayo. It remained airworthy until late 1941 when it was withdrawn from service and its structure, along a complete engine, donated to the city´s *Escuela de Artes y Oficios*[3] where it was used as an instructional airframe for mechanics. The ultimate fate of this airframe is unknown.

3 School of Arts and Occupations.

Nieuport-Delage NiD.121C.1

Developed in 1932 to compete in the contest to select the new AdA single-seat fighter, the Nieuport-Delage NiD.122/123 were a series of parasol type monoplanes of metallic construction which introduced a number of advanced features for their time such as a fire suppression system, complete radio transmitter/receiver equipment and pneumatic brakes for the landing gear wheels. The most innovative feature was, however, also the most peculiar feature, the engine cooling system which used a radiator built into the wing, where air was sucked in through slots in the leading edge of the wing and expelled through the trailing edge. Not even these advanced details weighted in favor of the model and in the end the AdA technical evaluation committee selected the Dewoitine D.500 instead, ruling the Nieuport NiD.122 model as underpowered, having a weak undercarriage and its radiator to be prone to damage in combat as well as inefficient when the aircraft was positioned at certain angles of attack. In order to compensate Nieuport for the prototype development expenses the French government authorized this model to be offered to foreign customers and in the meantime, Peru, was desperately looking for fighters in Europe in the face of the impossibility to obtain modern aircraft from US-based companies, CAP´s natural military hardware suppliers at the time. Therefore, in early 1933 Mr. Paul Dietrich approached the Peruvian *MG* offering – amongst other types – the Nieuport-Delage NiD.121C.1 as a fast, modern and powerful fighter capable of out-running and out-climbing the Curtiss Model 35-A "Hawk II", Colombian *Aviación Militar´s* main fighter at that time. With few options available and even with this machine still in the prototype status, the Peruvians reached a pre-agreement and committed to buy a batch of 10 machines, on the premise that the following modifications were implemented:

- Fitting of the 720 hp Lorraine-Dietrich 12 Hdrs *in lieu* of the original 550 hp Lorraine-Dietrich 12 Hars.
- Fitting of machine guns employing 7.65mm caliber ammunition[4].
- Fitting of under wing bomb racks for up to 120kg of bombs.
- Aircraft should be able to operate with floats as well as on wheels.

In order to not leave Peruvian skies undefended while the lengthy modifications and production process of the NiD.121 took place, Nieuport offered to sell three NiD.626 fighters – an export version of the NiD.62 series – but, despite looking favourably at the proposal in the beginning, the report issued by CAP´s technical committee considered the model unsuitable for use in the harsh environments of the Amazon forest, thanks to its wooden fuselage construction, and the proposal was rejected.

Although the Peruvian delegation had taken a first look at the NiD.121C.1 prototype as early as March 1933, it was only on September of that year that the series NiD.121C.1 was finally shown to Peruvian officers during their visit to the Nieuport factory in Villacoublay. This model, however, was powered by the original 550 hp Lorraine-Dietrich 12Hars engine, as the more powerful 12Hdrs was still in the development stage. This fact didn't prevent the Peruvian authorities confirming the previous order for ten machines shortly after the evaluation. The improved NiD.123 with the 720 hp Lorraine-Dietrich 12Hdrs inline engine was shown to the commission on July 18, 1934 but its flight characteristics were deemed unsatisfactory and the upgraded model was discarded and the order proceeded with the original ten NiD.121s. As in the meantime of these events the proposed sale of a trio of the readily available NiD.626 fighters fell through, the CAP ordered an additional two NiD.121C.1 fighters, for a total of 12 units. After almost two years from the signature of the original contract Nieuport-Delage began with the deliveries of these aircraft in December 1934, completing the CAP order in February 1935.

In service

The arrival of these fighters took place right after the ratification of the Salomon-Lozano treaty which brought peace between Colombia and Peru and put an end to the planned deployment of the new fighters in the Amazon forest. Five machines, however, were briefly employed by the *4to*

4 Standard ammunition caliber employed by Peruvian armed forces.

Escuadrón Mixto from "Alferez Huguet" seaplane base at Ancon, unit subordinated to the *Tercer Escuadrón de Aviación* (Third Aviation Squadron – III EA), where new crews began conversion to the type. It was during this period that the first accident of the type took place when, on May 9, 1935, a fighter identified with number 4-C-5 suffered a landing mishap flipping over its nose while taxiing on the beach contiguous to the seaplane base, sustaining moderate damage to its tail, engine and propeller, repairs that kept the aircraft grounded until early 1937. The 4 EM existence was short lived and by June 1935 the unit was disbanded and its aircraft transferred to Las Palmas AB where, along eight newly delivered Nieuport fighters, formed the *2 Escuadrón de Caza* (Second Fighter Squadron, 2 EC), receiving the correlative CAP numbers II/2-C-1C to II/2-C-12C.

The NiD.121 speed and rate of climb were appreciated by experienced pilots, although novice pilots found its controls to be unforgiving, especially during landing. On July 9, 1936, Nieuport II/2-C-3C was heavily damaged after it flipped over during landing. However, the first total loss of one of these machines took place on October 21, 1936 when 2nd Lt. Alfredo Silva Boggiano was killed when he crashed his fighter against the roof of one of the hangars in Las Palmas AB while performing a mock combat against a Caproni Ca.114 aboard Nieuport number II/2-C-7C. The falling debris caused by the impact also caused heavy injuries to a pair of mechanics which were inside the hangar giving maintenance to a Vought O2E-3B "Corsair".

In January 1937, in accordance with the Air Force Strategic Distribution Plan designed the previous year, nine *Newports*[5] from 2 EC were flown to "Capitan Guillermo Protzel del Castillo" AB in Vitor, Arequipa, subordinated to the Second Air Region. The limited range of these aircraft presented problems for the planned deployment of the unit to its new base and, after weighing up several alternatives, it was decided to use the aircraft floats for the last time and with that purpose the fighters were sent in groups of three to Ancon where floats were fitted. After that the amphibious fighters took off from the bay and headed to Mollendo, making refueling stops at Pisco and Marcona, in the coastline. In Mollendo the aircraft were loaded into flatbeds and carried to Vitor by railroad. The unit frequently deployed one or two of its three sections to the forward airfields in Juliaca, Tacna or Mollendo as part of their training and border protection duties.

Towards the end of 1938 the lack of spare parts, specially for the Lorraine-Dietrich engine, caused by the logistics problems in France had become a critical factor in the decline of the NiD.121 fleet operations, and this situation reached its peak in early 1939 with the complete grounding of the Nieuport fleet due to lack of spare parts. Facing the impossibility to secure vital spares for the continuation of the type operations the Peruvian CGA took the decision to withdrawn the NiD.121C from service, with all surviving aircraft sold as scrap by October 1940.

5 Name given to them by Peruvian crews.

Series of profile as well as ¾ views of Nieuport-Delage NiD.121C.1 prototype taken at the Nieuport factory in Issy-les-Moulineaux. (Liebert)

A NiD.121C.1 series aircraft fitted with floats during trials at Sartrouville. (Liebert)

The NiD.123C.1 demonstration aircraft seen on the Seine River during trials at Sartrouville in October 1934. Note Peruvian colors on vertical surfaces. (Ledet)

A NiD.121C.1 taxies after landing at Las Palmas in 1936.

Ancon, May 1935. NiD.121's narrow landing gear was the cause of several incidents during the type operational life. Here, a Nieuport from 4[to] Escuadron Mixto *identified with the number 4-C-5 receives attention by "Alferez Huguet" seaplane base mechanics after it flipped over during a landing on the beach. (Currarino)*

This Nieuport Nid.121 identified with CAP number II/2-C-3C and factory number 7 suffered a mishap during landing, sustaining heavy damage. Fortunately its pilot wasn't injured, but the machine spent several months in the repair shop. Note the bomb racks under the wings. (Currarino)

A CAP NiD.121C.1 identified with number II/2-C-4C take off from Las Palmas in 1936. (Gentili)

Below, right: Lineup of Newports from 2 EC seen during a ceremony at Las Palmas held on September 23, 1936. (Maguiña)

Below: 2nd Lt. Alfredo Salazar Southwell and his sibling portrayed in front of one NiD.121C.1 from 2 EC. Las Palmas, 1937. (IEHAP)

Below, right: Two CAP officers and an unidentified civilian pose with this NiD.121C.1 from 2nd fighter squadron, identified with registry II/ 2-C-4C. (Currarino)

MS.231 E/1A-1B, V Escuadrón de Instrucción, *Las Palmas AB in 1937. Camouflage: the aircraft was finished in* Gris Bleu Fonce *overall with bare aluminum metal panels. White numbers on fuselage and black manufacturer data on tail.*

NiD.121C.1, II/2-C-3C, 2do Escuadrón de Caza, II Escuadrón de Aviación, *Vítor airbase autumn 1937. Camouflage:* Vert Fonce *on top surfaces with bare aluminum metal lower fuselage and wing under surfaces. Black numbers on fuselage as well as on the aircraft manufacturing data vertical surfaces.*

Potez 390A.2, II/1-BO-2B, 1er Escuadrón de Bombardeo y Observación, II Escuadrón de Aviación, *Vitor airbase in 1937. Camouflage:* Vert Fonce *on top surfaces with bare aluminum metal lower fuselage and wing under surfaces. White number on fuselage and black aircraft manufacturing data on rudder. This aircraft was lost on September 14, 1937 during a training flight taking the life of 2*[nd] Lt. *Alfredo Salazar Southwell in the process.*

Potez 390A.2 fitted with floats as seen during its trials at Sartrouville during the summer of 1934. Camouflage: Vert Fonce *on top surfaces with bare aluminum metal lower fuselage and wing under surfaces. Natural metal floats.*

CHAPTER THREE
THE DELTAS
DASSAULT-MIRAGE 5P/5DP
IN PERUVIAN SERVICE

By the mid sixties the *Fuerza Aérea del Perú* (Peruvian Air Force – FAP) fighter arm operated 16 Hawker F.52 Hunter, 18 North American F-86F Sabre and a similar quantity of Lockheed F-80C Shooting Star jets, plus about forty Republic F-47D Thunderbolts. While the before mentioned jets had made their entrance into service with FAP during the middle of the past decade, the piston engine Thunderbolts arrived in Peru through the several incarnations of the US defense aid programs launched since the late forties and their operational life was rapidly approaching the end. As consequence the FAP high command prepared and launched, in early 1967, a fighter competition looking for a state of the art jet fighter bomber to replace both the F-80C and the F-47 aircraft. On February 26, 1967, the air force High Command ordered the creation of a technical intelligence group with the aim to travel to France in order to evaluate the latest technology in French combat aircraft. The group, led by Gen. Gilardi, arrived in France a few weeks after and performed a very detailed inspection and evaluation of the Mirage IIIC fighter bomber, after which the officers were left with a very favorable impression of the aircraft. This visit was followed shortly afterwards by the visit of the *ad hoc* purchase commission, composed by Maj. Gen. Schroth and three officer pilots whom, between March 2 and 3, visited the Dijon and Cazaux airbases, flying both the Mirage IIIC and B. Impressed by the report submitted by the technical officers the Peruvian authorities requested its French counterparts to immediately transfer between three and six Mirage IIIC fighters, which was opposed by the AdA as it was the middle of its own rearming process. Frustrated but not discouraged, the Peruvian officers were offered the new Mirage 5J which, while it also fulfilled FAP requirements, was cheaper and readily available.

Peruvian officers pose in front a Mirage IIIC at Mont de Marsan airbase, March 1968. (Garcia Falkenheiner)

Prototype for Mirage 5P shortly after its first flight. (Mejia).

After completing the additional test and evaluation process the FAP ruled in favor of the Mirage 5, discarding the contending Northrop F-5A "Tiger" offered by the United States. Negotiations took place and an agreement was signed between the Peruvian and French governments for the purchase of twelve machines, training of six pilots and 30 mechanics, as well as the introduction of a number of modifications to the Peruvian aircraft. Identified as Mirage 5P (for Pèrou), the aircraft differed from the standard Mirage 5F (denomination given to the Mirage 5J in French service) on the fitting (or absence) of the following items:

- A Marconi AD370 B radio compass replacing the original manufactured by CSF;
- A VOR-ILS marker Marconi 6401 in lieu of the one manufactured by SOCRAT;
- CSF CC 262 and a SARAM TRAP 28 VHF radio systems replacing the 2 TRT UHF;
- No TACAN
- A Nord 6071 B Tele-command system;
- No radar;
- A SFIM 154-5 central replacing the original Sperry inertial platform;
- Capacity to carry a 1100 liter or an AS.30 missile on the ventral fuselage rack.

A factory fresh Mirage 5P bearing number FAP 183 rolls on the runway for its first flight. The first batches of Mirage 5P delivered to the FAP wore a standard Armee de L'Air two tone Grey/Green camouflage. (Liebert)

A rear view of Mirage 5P number 182 at the tarmac of Las Palmas airbase, shortly after its arrival from Chiclayo on July, 1968. (IEHAP)

Col. Boluarte stands next to Mirage 5P number 182 shortly after arrival to Lima. (DIRIN)

Le Bourget, June 1969. A factory fresh Mirage 5P identified with FAP number 185 is exhibited in the Avions Marcel Dassault-Breguet Aviation section at the 1969 Air and Space fair held on that location. Note the RPK-10 dual fuel/bomb carrier system under the wings. (Liebert via Nuñez Padin)

The resulting aircraft, albeit a rather simplified version of the originally desired Mirage IIIC, was capable enough to cope with the requirements of a territory with similar meteorological conditions than those found in Israel, the first user of the Mirage 5.

On July 20, 1967, Peruvian State and Dassault representatives signed the *MARTILLO I* contract for the purchase of twelve Dassault-Mirage 5P and, by the end of that year, the first production

Mirage 5DP number FAP 197 taxies for its test flight at the Avions Marcel Dassault company factory in Bordeaux – Mérignac. (IEHAP)

machine rolled out of the production line. In the meantime the Peruvian *Ministerio de Aeronáutica* (Air Ministry) had ordered the assignment of half a dozen of the FAP most experienced fighter pilots to travel to France in early July in order to receive instruction on the type. After arrival to France the select group traveled to Dijon-Longvic airfield, home base of the famous 1/2 *Cigognes* fighter squadron, where they received first-hand technical training on the aircraft's systems and flight procedures. After completing the ground lessons, the Peruvian aviators moved to Luxeuil airfield to receive training on the Mirage IIIC simulator located there and then moved to Mont-de-Marsan airbase, in southern France, where they began actual training flights on the Mirage IIIC, as the Peruvian Mirage 5P aircraft were still being assembled on the production line. Once at Mont-de-Marsan, Peruvian pilots carried out multiple sub and supersonic flights, as well as zero visibility landing approaches, instrument flights, amongst others. Air-to-air combat training followed, using F-100 "Super Sabre" fighters, based on the nearby Cazaux airbase, as opponents. Finally, Peruvian pilots completed their training on air-to-ground combat procedures, employing rockets and live bombs on the airbase's training ground located nearby.

The arrival of the first Mirages and the creation of *Escuadrón Aéreo* 611

The Peruvians requested that the first aircraft to be delivered as quickly as possible and therefore the *Armée de l'Air* agreed to deliver the first two aircraft (numbers FAP 182 and FAP 183) by air, and the disassembled airframes were loaded into a pair of Transall C-160 transports and flown to the south American country, arriving in Peru on June 15, 1968, after following the Bordeaux–Merignac–Dakar–Recife–Belem–Manaus–Chiclayo route[1]. Assembly and testing of the aircraft was completed on July 16 when the aircraft were officially handed over to *Escuadrón Aéreo 611* (Aerial Squadron 61 – EA 611), which entered in operations with six pilots, two of which already qualified as instructors. This unit was part of *Grupo Aéreo Mixto* (Mixed Air Group) N° 13, name changed shortly after to *Grupo Aéreo* (Air Group, GA) N° 6 operating from "Capitan Jose Abelardo Quiñones Gonzales"[2] AB. The first "Mirachista" group, as Peruvian air force Mirage pilots were christened, was composed by the following officers:

- Cmdr. FAP José Guerra Lorenzetti, EA 411 commanding officer
- Maj. Augusto Romero Ferreccio
- Capt. Gonzalo Luzza

1 The French seized the opportunity to demonstrate the capabilities of the twin engine transport, flying from Lima to Cusco, located at a height of 11 000 ft, where it landed on a semi prepared field.

2 Previously known as "Teniente Coronel Pedro Ruiz Gallo", this airbase was renamed in honor of the military aviation hero on May 10, 1966.

- Capt. Arnaldo Velarde Ramírez
- Capt. Juan Burga Bartra
- Capt. Cesar Faura
- Lt. Hernán López

On July 19, 1968, Majors Augusto Romero Ferreccio and the recently promoted Cesar Gonzalo Luzza flew Mirages N° 182 and 183 from "Capitan José Abelardo Quiñones Gonzales" AB to the "Jorge Chavez Dartnell" international airport, in Lima, breaking the sound barrier and reaching Mach 1.8 in the process. Four days later, these pilots took part in the official display and presentation of the FAP´s new aircraft during a ceremony held at Las Palmas AB on July 23, 1968.

EA 611 OPERATIONS

The demanding requirements of the Mirage 5P weapon system meant that only the finest, most capable flight and ground personnel were assigned to operate the system, turning EA 611 into a sort of elite unit within the FAP. The high grade of preparation of the human element tasked with the operation of the deltas became evidenced by its response during the several incidents that took place during the unit's daily operations. On account of their importance the following deserve to be mentioned:

1. On March 8, 1971, the Mirage fleet recorded its first loss, when Mirage Mirage 5P number FAP 188 crashed during a training mission. This incident reduced the number of operational aircraft to 13 machines. A few weeks later, on April 17, the EA 611 suffered the loss of its only twin seat aircraft when the recently delivered Mirage 5DP number FAP 197 had to be abandoned by its pilot over the ocean due to a human mistake.
2. On August 2, 1972, during a live ammunition training mission over the Reque gunnery range flown by Maj. Gonzalo Luzza (callsign "Impala") aboard Mirage 5P number FAP 194, a faulty 30mm round exploded right after leaving the barrel, the gasses from the explosion causing the engine to "flame out". Keeping his cool and willing to save the aircraft despite the fact that he was at only at a height of 1,600 ft (500 m), Luzza radioed the emergency call to base control tower and at the same time, followed the established procedure in case of an engine malfunction:

- Engine re-start,
- Level the aircraft with the horizon;
- Apply full throttle.

The standard procedure during an engine malfunction was either to proceed with engine restart – if enough altitude was available to complete the checklist – or to abandon the aircraft immediately by employing the ejection seat. Fortunately for the pilot, the RPM indicator livened up as his

Dassault test pilot Tixador prepares to take this factory fresh Mirage 5P to the air for its first flight from the runway at Mérignac on May 25, 1969. This aircraft will be assigned with the number FAP 186. (Dassault Aviation via Liebert)

Mirage FAP N° 182 flew over typical scenery of the so called "volcano valley", located in Arequipa, south of Peru. (IEHAP)

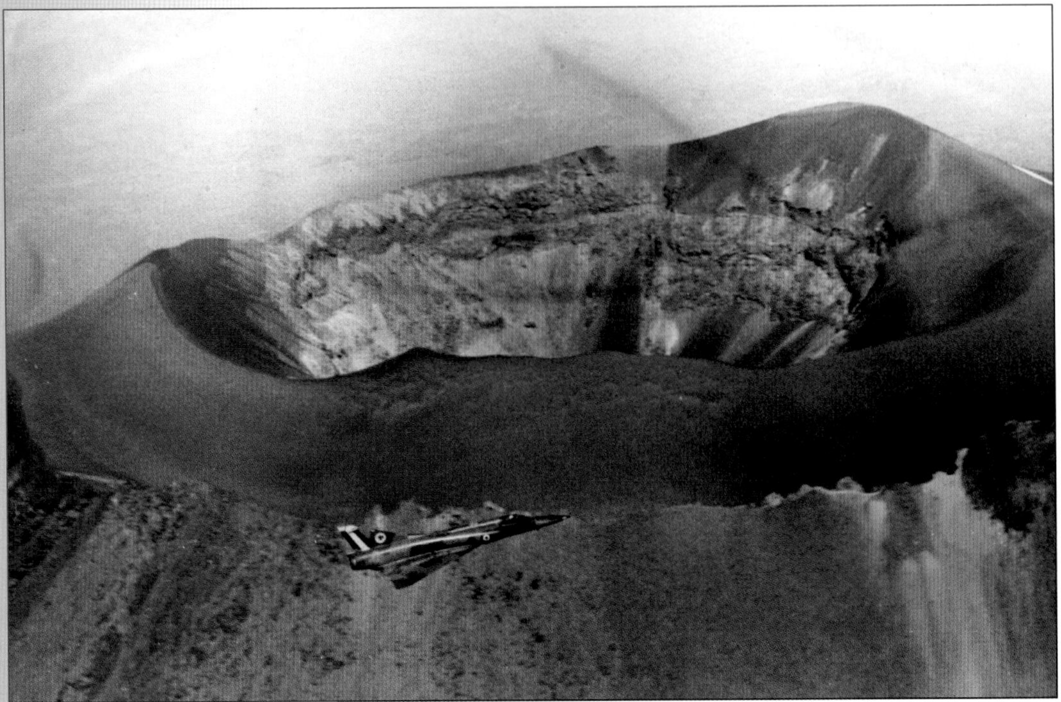

aircraft's engine came back to life, and after checking proper operation of the other vital systems, he contacted the base to inform about the situation, requesting landing clearance. In the meantime, a still nervous Luzza resorted to an unorthodox procedure to calm his fear by punching himself hard in the genitals to make the pain drive away the nerves and help him focus on his tasks. Apparently that worked and moments later, he flawlessly landed his aircraft back at the runway of "Capitan Jose Abelardo Quiñones Gonzales" AB, where he was warmly received by the emergency crew awaiting there. Immediately after this incident, EA 611 Command forbade the use of the type of ammunition involved in the incident until the board of inquiry issued its report, arriving at the conclusion that the premature explosion was caused by a defective fuse.

3. A day after the incident involving Mirage number FAP 194, the EA 611 deployed a four-ship formation on a training flight to "Capitan Guillermo Concha Iberico" AB in Piura, where a fly-past was performed over the GA N° 7 officers tribune as part of the ceremony held at that location. After completing its first passage, aircraft number FAP 196, flown by Lt. Hernán Ampuero, was hit in the frontal fuselage by a bird which opened a hole in the nose cone, covering the windshield with blood and other remains. Fortunately, the impact did not cause any damage to the aircraft's vital systems and, after checking his own integrity for any injuries, Ampuero, following Maj. Luzza's example, headed towards Chiclayo guided by Capt. Carlos Vasquez Zegarra, the 4th pilot in the formation. After expending the remaining fuel circling over Chiclayo, both pilots readied to land their aircraft as Vasquez maneuvered his Mirage to the lead position and over Ampuero´s, who carried out his landing without any problems, stopping barely a few meters away from the emergency arrestor barrier. It is necessary to highlight that the emergency landings performed by both Luzza and Ampuero were carried out just as the airbase landing strip was being subjected to maintenance, and therefore its whole extension wasn't available for operations.

The Nord Aviation AS.30 Air to Surface Missile

Among the plethora of ordnance augmenting the FAP inventory with the arrival of the Mirages were a total of 35 Nord Aviation (later Aerospatiale) AS.30 air-to-ground missile, which was a development from the earlier AA.20, an air-to-air weapon. The AS.30 was powered by a two-stage

A Mirage 5P displaying part of its available ordinance. Note the AS.30 missile in the middle, as well as the Matra JL-100 rocket launcher/auxiliary fuel tank combination. (IEHAP)

solid fuel rocket engine, measuring 3.7 m in length with a diameter of 340 mm. The weapon´s warhead weighted 240 kg (529 lb) out of a total of 520 kg and was composed of impact-fused Semi-Armor-Piercing High-Explosive, or SAPHE. Like its predecessor, the AS.30 had a MCLOS[3] guidance system, which basically constituted World War 2 technology. Operating this weapon was far from simple as the pilot, after releasing the missile, had to track its path continuously by means of a flare located at the rear end of the missile, manually steering the small, fast traveling weapon towards its designated target via a joystick interface in the cockpit – all this while trying to keep control of his own aircraft. Needless to say, it was a very complicated weapon to guide to its target and optimum results could only be expected when deployed by experienced crews.

The Peruvian Air Force operated rather large numbers of AS.30 ever since the mid-sixties, when this weapon arrived as standard ordnance for the Canberra bombers operated at the time by EB 622. With the arrival of the Mirage, a second vector was added for the weapon, and training on its operation was included as part of EA 611´s *Programa de Entrenamiento* or Operational Training Program (PDE) for the year 1971. It was during that period that Maj. Luza – already chief of EA 611 at that time – performed the first launch of an AS.30 missile from a FAP Mirage 5P. On May 12, 1971 during the PDE 1971, a Canberra from EB 622 and a Mirage from EA 611 flown by Maj. Luzza, each armed with an AS.20[4] missile, were to launch their weapons against a target located at the Reque gunnery range; the most accurate hit would then earn the corresponding pilot the chance to land, rearm his aircraft with a live AS.30 and perform a launch. Maj. Luzza, aboard Mirage N° 186, scored the most accurate hit and, consequently, was authorized by the chief of the *Comando de Operaciones* or Operations Command (COMOP) to return to Chiclayo, rearm and perform a second launch. Thirty minutes later, Maj. Luzza's Mirage fired its AS.30 over the range, scoring a perfect hit.

To improve the capabilities of EA 611 crews, the Peruvian AM and Dassault Aviation signed an agreement for the arrival of a technical advisor to help improve the unit´s use and employment of the Mirage weapons system. Therefore, on June 18, 1975, *Sergent-chef* Michel Kolly arrived in Lima, becoming immediately assigned as an advisor to the EA 611 High Staff. Kolly had graduated from the AdA's *Ecole de Chasse* at Tours in December 1969 and was, at the time of his deployment to South America, assigned to the Mirage IIIC-equipped *Escadrón de Chasse 1/2* based at Dijon

3 Manual Command to Line of Sight.

4 AS.30 predecessor, armed with a 33kg warhead.

Las Palmas, July 23, 1971. A Mirage 5P pulls over this airbase as part of the aerial display held during the military aviation day. Below, a Hawker Hunter F.52 and another Mirage 5P display their entire combat ordinance. (IEHAP)

AFB, unit which he had joined on January 4, 1971. A display pilot himself, besides a colorful character, Kolly usually flew a Pitts Special biplane from Lima and Chiclayo. Unfortunately, in February 1981 during one of such flights, the French aviator was killed when his Pitts Special plunged into the sea in front of Pimentel harbor.

New requirements, originating both as a result of operational experiences as well as strategic planning designed in accordance with the geopolitical changes in the region[5], led to the need to improve some of the Mirage 5P avionics and systems. Accordingly, in the year 1974, the FAP HQ designed a project aimed to improve the Mirage 5P navigation and identification systems by fitting the following systems:

- Litton LN-3 Inertial Navigation System;
- EMD 3300E Identify Friend or Foe-IFF – system, replacing the original LMT 3560J

The project was executed by the SEMAN (Air Force Maintenance Service) with assistance from AMD-BA and completed progressively on the Mirage 5P fleet over a two-year work period. By installing modern radio altimeter and inertial platforms, the upgrade improved the Mirage 5P navigational capabilities by enabling the pilot to identify the aircraft`s position by means of an instrument in the cockpit. During this limited upgrade process works at SEMAN the aircraft received a new paint scheme consisting of sand and dark earth over light grey undersurfaces.

By 1976 and thanks to the acquisition of additional airframes as well as replacements for the airframes lost during routine operations, the FAP was able to field a total of 36 Mirage 5P3, enough to form an additional Mirage squadron, called *Escuadrón de Aviación 612* "Los Gallos" *(Fighting Gamecocks)* or EA 612. For logistical reasons in the beginning the newly activated unit operated alongside its sister unit from "Capitan Jose Abelardo Quiñones Gonzales" AB in Chiclayo but was later deployed to La Joya AB, in Arequipa, a year later.

Accidents, however, continued to shadow the Mirage 5P operational life. On March 22, 1977, a Mirage 5P3 identified with number FAP 101 crashed while on the course of a training flight, fortunately without human losses as its pilot managed to abandon the aircraft using his ejection seat.

Lack of air-to-air weapons in the Mirage inventory had limited the efficacy of the Mirage 5P as an interceptor ever since its entrance into FAP service. Since the purchase of French or US missiles was out of question due to either financial or political reasons, the FAP Mirage fleet had been

5 Like the coup d'état launched in neighboring Chile in the year 1973 by the military leadership against the government of President Salvador Allende.

Tail from a Mirage undergoing major repair at Avions Marcel Dassault plant in Bordeaux in 1974. Note the fighting gamecock insignia on the tail, emblem of EA 611. (Liebert)

restricted to operate only with cannons in the air-to-air role. This situation changed in 1977 with the introduction of the Su-22A Fitter F fighter-bombers and their associated weaponry, which also meant the arrival of substantial numbers of Vympel R-3S (coded AA-2A "Atoll" by NATO) air-to-air missiles. Since the R-3S was essentially a reverse-engineered copy of the AIM-9B Sidewinder – a weapon integrated to the original Mirage 5J from the very beginning – the Air Force HQ ordered SEMAN to perform all the required modifications to the Mirage airframes in order to successfully carry and fire the missile, a project undertaken between 1977 and 1979. In the meantime, operations continued by both Mirage units, and with these also came the unavoidable operational incidents. On March 22, 1977, EA 612 suffered the loss of Mirage 5P3 number FAP 101 which suffered an engine malfunction during a routine flight, forcing its pilot to use its bang seat. In a twist of events the aircraft, instead of diving to earth towards its end, glided to the surface and gently touched ground causing only minor damage to its lower structure.

The 1978 PDE designed for both EA 611 and EA 612 included the so-called "Operational Efficiency and Capability Training Program" exercise, which was designed to improve the operational capabilities of the *Mirachistas*. According to this, each pilot was tasked with a training mission, consisting of a low altitude navigation sortie covering no less than five waypoints, with photographic checks at each point, followed by bomb, rocket and cannon practice runs at the Reque gunnery range, where instructors were to be present to evaluate their accuracy. Meanwhile, the exercise also tested the work of ground crews, where maintenance and flight configuration response times undertaken by personnel assigned to each aircraft were also evaluated. Furthermore, nocturnal training sorties, identified as "Murciélago" – or Bat, in English – as well as deployments to airfields in the eastern Amazon rainforest and southern areas of the country were also included in the exercise, as part of the training process for ground and air crews.

Foreign visitors

The high degree of expertise on the Mirage acquired by the EA 611 and 612 was recognized by others of the type's operators in the region and a number of FAP officers from these two units shared their experiences with members from other regional air forces. The first to show interest was the *Força Aérea Brasileira* – Brazilian Air Force or FAB – which invited a number of FAP officers to Brazil between 1978 and 1980 in order to share knowledge on the Mirage weapons system, a gesture returned a few years later when the FAP issued Brazilian Air Force Lt. Col. Alfredo Severo

Luzardo an invitation to take part in the tactical-operational training held by EA 611 at Chiclayo during 1982. *Fuerza Aérea Argentina* – Argentinean Air Force or FAA – officers also visited Peru to train with EA 611 when a group of pilots from the *VIII Brigada Aérea* (VIII Air Brigade) visited GA N° 6 facilities in Chiclayo during the year 1983 to train and share experiences on the Mirage with their Peruvian counterparts. Finally, in late 1985, a group of pilot officers from the *Fuerza Aérea Venezolana* – Venezuelan Air Force or FAV – also arrived in Chiclayo to train with the EA 611.

Mirages over "Falso Paquisha"

The Mirages took part in the short but violent border clashes between Ecuador and Peru – also known as the "Falso Paquisha" (Fake Paquisha) conflict – during the early part of 1981. This conflict began during the early hours of January 22, 1981 when an *Aviación del Ejército* – Peruvian Army (AEP) Aviation – Mi-8T transport helicopter was hit by small arms fire coming from troops from *Ejército del Ecuador* (Ecuadorian Army – EdE) forces while overflying a Peruvian border post located within the "Cordillera del Cóndor", a mountain range acting as the natural border between the two countries. Shortly after this incident, an EP platoon performing a routine border patrol was attacked by EdE forces near one of the surveillance posts inside Peruvian territory, evidencing the infiltration of foreign forces. Two days later and after a series of urgent meetings and deliberations, the Peruvian Government, led by President Fernando Belaunde Terry, requested the Air Force to perform a reconnaissance over the compromised area with the purpose of obtaining concrete proof of Ecuadorian infiltration before ordering the armed forces to step in. To accomplish that requirement, on the morning of January 25, a Gates Learjet 25B from *Escuadron Aerofotográfico 331* (331 Aerial Photography Squadron – EA 331) – took off from Las Palmas AB and headed north to Chiclayo, where it rendezvoused with a pair of deltas from EA 612 which escorted the unarmed aircraft to the Condor Mountain Range. The reconnaissance aircraft and its escorts arrived over the Comainas area at 1230 hours and, with good weather over the zone, it was possible to obtain good images that, once brought back to Lima, were used as proof to confirm the presence of Ecuadorian forces occupying former EP border posts located along and around the Comainas River, inside the Peruvian side of the Condor mountain range – posts that had been lying unoccupied by the EP since the late seventies[6]. With the new evidence, the Peruvian Government issued a diplomatic complaint note to the Ecuadorian Government, followed by an accusation in front of the United Nations demanding the immediate withdrawal of Ecuadorian troops from its territory. The crisis escalated when the Ecuadorian Government rejected the Peruvian accusations, despite the hard proofs, claiming that the alleged occupied positions were in fact the Ecuadorian populations of Paquisha, Mayaico and Machinaza, and that the incidents were caused by intruder EP patrols. This situation prompted the issuing of an ultimatum by the Peruvian Government to its Ecuadorian counterpart requesting the withdrawal of its forces from the occupied positions before January 26. With both sides displaying signs of intransigency and making a diplomatic exit to the crisis less likely by each passing hour, the government in Lima requested all the branches of the armed forces to plan a two-stage operation, code-named *Operación Grau*, composed of:

1. Limited zone operations aimed to expel the intruders, scheduled to begin on the morning of January 28, 1981.
2. A comprehensive series of offensive operations against airbases, oil refineries and other strategic targets inside Ecuador, in the event that the border operations developed into a full scale war and aimed to cripple its ability to conduct a protracted conflict.

The aerial operations part of this plan received the code name *Operación Rechazo* (Operation Rejection), and its execution was put under the command of the chief of *Ala Aérea N° 1* (1st Air Wing – AA N° 1), which grouped under a single command structure the several FAP units stationed

6 After an agreement between Ecuadorian and Peruvian governments to defuse the border crisis, caused by troop clashes inside a border sector that remained unmarked since the fifties.

Series of pictures depicting EA 611 operations from Chiclayo during the year 1974. (IEHAP)

Mirage 5P FAP 186 seen at Las Palmas shortly after delivery. The aircraft was finished on a standard French tactical camouflage of dark green/dark grey over light grey. (DIRAF)

north of the country[7], issuing a series of directives destined to ensure proper operation of the units in the face of the upcoming operations.

Regarding the Mirage units, the EA 611 received the order to remain "on alert", awaiting the call to action if the conflict worsened, as its crews received specific sets of orders including targets inside Ecuadorian territory. Meanwhile, EA 612 was to step into the first phase of the conflict with its crews receiving orders to fly CAP[8] or TARCAP[9] missions over the operations zone starting at 0500 hours of January 28. These missions were aimed to deny the *Fuerza Aérea Ecuatoriana* (Ecuadorian Air Force, FAE) the use of the airspace over the area of operations and to provide a safe operational environment for the helicopters from GA N° 3, which were to ferry the troops tasked with the recovery of the border posts.

With orders arriving from FAP HQ in Lima, GA N° 6 units began the planning of their respective operations which, once completed, were issued to COMOP in order to perform the required coordination between the different ground and air assets assigned to the theater.

With EA 611 in alert awaiting further orders, EA 612 was the unit selected to carry the burden of the Mirage operations during the conflict. The unit was given the responsibility of providing an umbrella for the A-37 and Su-22A from EA 712 and EA 111, respectively, which would perform strike operations against Ecuadorian positions, as well as for the EP and FAP helicopters that would ferry assault troops directly into the occupied positions. The unit´s command, in strict compliance with COMOP directives, issued a series of orders directed to ground and flight crews aimed to increase the unit´s availability in the face of the upcoming requirements, managing to put 12 Mirage 5P3 back in combat-ready condition. Similar measures were undertaken by EA 611's high staff, whose aircraft were readied and sent to dispersal area on a "24-hour alert" condition, with pilots awaiting in their cockpits the signal to take off to their previously designated targets inside Ecuadorian territory. Both units achieved a 93% availability level thanks to the efforts and high degree of preparation displayed by personnel from 606 Maintenance Squadron.

At 0615 hours of January 28, the calm over the Condor mountain range was broken by the sound of the powerful Atar engines as the first "Dardos" from EA 612 made their appearance over the area of operations on CAP duties. The first two aircraft roared through the skies flying 10,000ft over the top of the Condor Mountain Range, their established CAP altitude, leading a pair of A-37s Dragonflies from EA 712. Operations continued and at 1615 hours, a pair of Dragonflies, FAP 118 and FAP 125 flown by Lt. Calle and 2nd Lt. Maldonado, arrived in the area to attack Ecuadorian

7 It was composed by the 6, 7 and 11 Air Groups.

8 Combat Air Patrol.

9 Target Combat Air Patrol.

positions in and around PV-22. However, as the leader began his fifth attack run, a pair of FAE A-37B appeared in the area and engaged him in combat. The Ecuadorian aircraft, FAE 384 and FAE 381, flown by Capt. Rommel Romo and Lt. Victor Pena, respectively, gained a favorable position over the leading FAP Dragonfly, forcing Lt. Calle to quickly drop his nearly empty LAU-32 containers to take evasive actions as the FAE Dragonfly flown by Capt. Romo fired his rockets (!) and then 7.62 rounds against his aircraft. At the same time, Lt. Maldonado also entered the *meleé* engaging the aircraft flown by Lt. Pena. During the vicious dogfight that ensued, lasting seven minutes, both elements performed risky maneuvers at nearly tree top level; in the end, Lt. Calle managed to shake off his pursuer and disengage. The worsening weather conditions over the area, as well as an empty GAU-2B/A, forced Capt. Romo to give up and the Ecuadorian aircraft broke off and returned to their territory. Meanwile Lt. Calle, with his fuel indicator nearing Bingo[10], returned to Piura on one engine, landing safely. After debriefing, Lt. Calle's aircraft was inspected by ground personnel who identified several bullet holes on the starboard wing. Maldonado, who also returned to base on his own, claimed one FAE A-37 as damaged[11]. It has to be noted that a pair of Mirages from EA 612 were on the area on CAP patrol but their pilots, despite being fully aware of the events via the radio chatter, became unable to provide assistance to their colleagues from EA 712 due to bad weather and the terrain conditions which presented a risky environment for their operations. This incident also evidenced one of the major setbacks faced by the FAP combat units throughout the conflict as they were strictly forbidden to engage hostile aircraft across the Ecuadorian border thanks to an order issued by the Peruvian President himself – a directive that effectively hampered the capabilities of the Air Force to perform its duties effectively throughout the campaign. Nevertheless, operations continued and on January 28, the EA 612 launched a total of 40 sorties, divided between CAP and escort, culminating in a total of 60 flight hours. The unit´s ground personnel performed flawlessly and were able to put a "Dardo" back in the air within 15 minutes after landing. The efficiency displayed by GA N° 6 flight and ground crews allowed the EA 612 to have four machines in the air over the area of operations at all times, with two deltas on the return leg at the same time while two others departed to the area from Chiclayo. Finally, it is worth to note that the standard payload configuration for the Mirages on CAP missions was 2 x Vympel R-3S A-A missiles and 2 x 30 mm DEFA cannons, with 125 rounds each.

10 Minimum fuel state needed for an aircraft to return to base (aka "recovery").

11 Lt. Pena's aircraft was indeed damaged by a single 7.62 bullet on its port wing, which compromised the spar. *Historia Ilustrada de la FAE*, October 1999, page 118.

A nice aerial shot of Mirage 5P number FAP 189 fitted with 1300 liter drop tanks. (Bernard Régnier)

A Mirage 5P taxies in front of a lineup of Canberra bombers at "Las Palmas" AB during a ceremony held for the anniversary of Peruvian Military Aviation on July 23, 1973. (IEHAP)

January 29 was also a busy day for EA 612 with a total of 30 sorties launched. During one CAP sortie over the operations zone providing cover for the A-37Bs flown by Capt. Seabra, unit commander and Lt. Larrea, his wingman, flying aboard Cessnas FAP 116 and FAP 136, respectively, made their appearance over the area at 1205 hours. Due to poor weather conditions they had to be led into the area by Mirage 5P3 number FAP 191 flown by Capt. Ugarte, but upon arrival, they found their target covered by clouds and could not deliver their payload. In the meantime, Capt. Ugarte and his wingman departed the area after being relieved of their CAP duties by Maj. Guerrero and Capt. Hayes, flying Mirages FAP 108 and FAP 193, respectively. However, as Capt. Seabra and his wingman were abandoning the area, the FAC reported on the radio about an incoming patrol of FAE Dragonflies appearing over PV-22, apparently on a search and destroy mission against FAP and AE helicopters operating in the vicinities. Once again, bad weather hampered FAE operations as well, as the Ecuadorian Dragonflies quickly withdrew to their own territory when they found the area covered in clouds. This prevented, once again, interception by FAP Mirages which could not get near them due to the excessive CAP altitude they were flying at.

CAP sorties continued and at 1215 hours, EA 612 was back in the air providing cover for another wave of A-37Bs. This time, Dragonflies numbers FAP 118 and FAP 125, flown and Lt. Orihuela and 2nd Lt. Maldonado, respectively, appeared over PV-22 finding slightly better weather conditions than the previous flight. This allowed Lt. Orihuela to perform a single bomb run dropping a single Mk 82 250 lb iron and an AN-M41 cluster fragmentation bomb over Ecuadorian positions before weather conditions deteriorated again, preventing further attacks. Eight minutes later, a third wave of attackers appeared over the area and successfully struck Ecuadorian positions, including AA emplacements with rockets after being positively directed by *Soplón* (or snitch in English), the FAC in the area. "Dragonflies" FAP 124 and FAP 149, flown by Maj. Klatic and 2nd Lt. Hundskoff, attacked the objectives with rockets and cluster bombs before returning to base without incidents. In the meantime, while providing CAP for the Cessnas below, a pair of Mirages from EA 612 received information from the FAC about the presence of three Ecuadorian helicopters over Peruvian soil, flying at top speed towards their own lines. The Mirage pilots were quickly guided into the area by *Soplón* for a GCI[12] on the intruders which at that time were about to cross the top of the moun-

12 Ground Controlled Intercept.

Formation of Mirage 5P from EA 611 flying over Chiclayo during the early seventies. Note the unit insignia on the leading aircraft identified with number FAP 190.

tain range (division line between the two countries). *Dardo* flight leader requested authorization to follow the intruders into Ecuadorian territory. COMOP, however, hesitated and the helicopters reached Ecuadorian airspace, after which the Mirages were ordered to disengage and return to base.

During the evening of January 29, EA 612 was requested to launch four aircraft on escort missions on behalf of EA 921 and EA 331 which had to perform familiarization and photo reconnaissance sorties over the area of operations, respectively, and at 1600 hours four Mirage 5P3 scrambled from Chiclayo and rendezvoused in the proximities with a BAC Canberra B.56 and a Gates 25B Learjet heading to the Condor mountain range. The crew from the Canberra, however, became disoriented due to bad weather and lost its assigned escort of two *Dardos*, entering Ecuadorian air space in the process. Once the crew realized their location they quickly turned back towards Peruvian airspace but failed to re-enter on its pre-established waypoint, situation registered by the Peruvian radar operators which prompted the scramble from Talara of a pair of *Flechas* (Arrows), code name assigned to the Su-22A Fitter F fighter bombers from EA 111. Fortunately for the Canberra crew, the flight leader chose to close-in on the suspected intruder in order to make a positive identification before opening fire, thus preventing a friendly fire incident by doing so. Meanwhile, at 1850 hours and while the pair of Mirage 5P3 assigned to provide escort to the Learjet were returning to Chiclayo after completing their assignment, one of the *Dardos* experienced a failure on its INS while on POPEYE[13], leading its pilot to deviate from its intended waypoint. Low on fuel and almost in darkness, the pilot managed to return to base and perform an emergency landing, causing minor damage to his aircraft after crashing it against the emergency landing barrier.

No sorties were flown by EA 612 during January 30, but operations resumed the next day when a *Dardo* element scrambled from Chiclayo with orders to intercept a boogey detected by the radar flying at high speed over Peruvian territory. The intruder, a FAE SEPECAT Jaguar Mk.1 on an electronic reconnaissance sortie, quickly abandoned the area probably alerted from the ground of the looming menace of interception. A second mission was flown later that day by an element of Mirages which were ordered to escort a Learjet from EA 331 to and from the operations zone. The aircraft carried none other than Peruvian President Fernando Belaunde Terry on a visit to "Falso

13 Flying in and out of clouds / area of reduced visibility.

Paquisha" (PV-22) where he announced to the press that Ecuadorian forces had been expelled from Peruvian territory, showing a large quantity of weapons and other equipment captured by Peruvian forces after they were left behind by retreating Ecuadorian troops. To counter any potential threat, two additional Mirage 5P3s were dispatched from Chiclayo to patrol between "Teniente Pinglo" and the operations area.

FAP offensive operations were suspended on January 31, but that did not mean the end of 6 AG involvements in the conflict as its units continued to perform daily patrols over the area until March 13, when the *Grau* operations plan was officially called off. From the beginning of the crisis on January 22 to the actual cease of operations on March 13, GA N° 6 pilots from both units performed 100 sorties for a total of 153 flight hours.

Mirages 5P to Argentina

On April 2, 1982, military forces from Argentina invaded the Falklands Islands (Islas Malvinas for the Argentineans) on a "re-occupation of its former territory" from the United Kingdom, as announced by the military Junta ruling the country. The illegal move was widely condemned by the international community and had as a consequence – among others – an arms embargo imposed on Argentina by the United Nations Security Council. To make matters worse, British reaction was swift and determined, and a Royal Navy Task Force was sent immediately with orders to retake the islands and return them to the Commonwealth. Shortly afterwards, the skies over the Falklands became the combat stage for Royal Navy and *Fuerza Aérea Argentina* aircraft – the latter's Mirage III, IAI Daggers and Douglas Skyhawks, although flown valiantly, proved no match for the "all-aspect" AIM-9L Sidewinder air-to-air missile of the RN Sea Harriers, a fact that was reflected on the loss of 19 Argentinean machines during the conflict.

To replenish its ranks after the losses sustained during the short but violent campaign, the FAA opted to exercise the rights contained on a previously signed contract with the Peruvian

A diamond formation composed by Mirages from EA 611. Chiclayo, 1972.

With the arrival of new Mirage 5Ps in early 1976 the FAP Mirage fleet finish was standardized in a new "desert" patterns which was applied to aircraft as they went into major maintenance. One of the newly arrivals was Mirage 5P number FAP 102 seen here at the AdA Cazaux AB on July 5, 1975. (Liebert)

Government for the transfer of ten Mirage 5P3 fighters for a total of 50 million US$ (or 5 million apiece). However, the author believes that clarification in necessary regarding this matter since, even up to this day, it is maintained that Peru "helped" Argentina by selling these aircraft amidst difficult times, when in fact all the Peruvian Government did was to honor a previous arrangement dated December 14, 1981, signed between the Argentinean and Peruvian governments aimed for the transfer of ten aircraft to the Argentinean Air Force. A memorandum of understanding had been signed four years earlier in 1978, and by virtue of it, the Argentineans signed a memorandum of understanding committing to the purchase of the ten aircraft, while the Peruvians committed to keep them in full operational condition until the delivery date, which was yet to be specified by the FAA according to the agreement. It should be noted that the aircraft belonged to the earliest batch of deltas received by the FAP and therefore, had the highest number of flying hours logged onto their airframes. Despite this fact, the airframes were in overall acceptable condition and with plenty of flying hours left. There was at least a pair of airworthy *Maras*[14] still in service with the FAA by 2015.

On July 5, 1982, the Mirages – already Argentinean property and sporting numbers 183, 185, 186, 102 to 107 and 188 – were ferried from La Joya AB to Jujuy in Argentina, and then to Tandil. Peruvian Air Force pilots flew the aircraft under complete radio silence over Bolivian airspace – which was powerless to intervene anyway – and reached the first stage of their destination successfully. After a short stop at this location, the formation reached Tandil the following day. Their arrival took place just a few days before the end of operations in the South Atlantic archipelago and therefore, no Mirage 5P took part in the war. It is important to note that, after the delivery of the deltas, the FAP performed a series of additional flights undertaken by a FAP C-130 carrying spare parts, 9K32M Strela-3 (SA-7 Grail according to NATO) surface-to-air missiles, the remaining stock of FAP´s Nord AS.20 and AS.30 missiles, as well as torpedoes. More importantly, these flights delivered an item probably more valuable to the Argentineans than the previously mentioned equipment: several sets of RP30 auxiliary fuel tanks, dozens of which were dropped by the FAA Mirages over the South Atlantic during the offensive operations against the British fleet around the islands.

As a side note, it is worth mentioning that the Peruvian Government did become involved in a triangular trade operation to procure Mirage IIICJ fighters, *Shafrir* missiles and 1,500 lt drop tanks from Israel, which could be considered as the real helping gesture from Peruvian President

14 Name given to the Mirage 5P in FAA service.

Fernando Belaunde Terry and his First Minister Manuel Ulloa towards Argentina. As the story goes, Belaunde and Ulloa authorized the Peruvian *Ministerio de Aeronáutica* (Ministry of Aeronautics, MA) to issue blank purchase orders as well as destination certificates, which were later filled in by Israeli companies. Thanks to the help from the Peruvian Government, Argentina was able to elude the embargo and lay their hands on millions of dollars' worth of military equipment that would have been impossible to procure otherwise. The deception was so elaborate that even some of the Mirage IIICJ received FAP camouflage paint schemes, markings and even fake numbers, and although British Intelligence eventually discovered the gambit, by that moment most of the equipment was already on Argentinean soil.

Proyecto Pegaso: the Mirage 5P4 Upgrade

On June 8, 1981, FAP and AMD-BA signed the *MARTILLO X* contract for the purchase of three Mirages, two single seat and a two seat aircraft, belonging to a new, upgraded variant identified as Mirage 5P4/DP-4. The new variant was a dedicated precision strike aircraft fitted with the following equipment:

Fully integration between NATO and Warsaw Pact weapons system, enabling the aircraft to employ both French and Soviet weaponry such as the Soviet-built *Vympel* R-3S "Atoll" and R-13M "Advanced Atoll" air-to-air missiles;

- Litton LN-33 Inertial Navigation System platform;
- TRT AHV6 radio altimeter;
- Thomson-CSF VE 110 CME Heads-Up Display replacing the CSF 97;
- Hands On Throttle-And-Stick (HOTAS),
- Thomson-CFM TMV 630 laser rangefinder/CSR laser air-to-ground telemeter fitted in a fairing located under the nose,
- EAS RNA 720 VOR-ILS;
- TVT TA340 IFF-TH
- 2 EAF UHF and VHF radios and
- In-Flight Refueling Probe (IFRP).

This agreement was followed by another contract, signed in March, 1982, for the complete upgrade of the 15 Mirages to this standard, work to be performed entirely in the SEMAN facilities in Lima. This deal included the conversion in France of an existing machine, FAP 110, which at the time was undergoing major repairs at Dassault BA facilities after it was damaged during an incident

After the loss of the original Mirage 5DP, FAP 197 in the early seventies, the Peruvian air force ordered a replacement aircraft which was delivered already finished in desert scheme.

FAP Specialist Sub Officers kneel for this picture in front of an impressive Mirage 5P3 formation. Note that the leading aircraft, still in original French grey/green camouflage, is armed with Soviet made Vympel R-3S Air to Air missiles, as well as FAB bombs are part of the aircraft ordinance, thanks to local modifications. (Ponte)

A side view from the same formation. Visible in the background is the Pitts Special acrobatic biplane usually flown by the French Michel Kolly. (Ponte)

in the year 1978. After completing its repairs FAP 110 was upgraded with the aforementioned equipment, with system tests performed at Istres between September 1985 and March 1986. This was the first Mirage III/5 ever to be fitted with an IFRP and tests showed that the aircraft´s general performance wasn´t affected by this modification and that the fuel transfer ratio allowed by the probe reached 500 l/min. Once the testing was completed the airframe was shipped back to Peru along with Dassault technical supervisors, the fifteen upgrade packages and a single three-point tanker refueling modification kit[15].

Once the conversion kits arrived in Lima, SEMAN started with the progressive conversion of the Mirage 5P3 from EA 611 to the P4 configuration on a process known as "Proyecto Pegaso" (or Pegasus Project). Budget restrictions, however, meant that the project took more time than originally intended and, accordingly, the last Mirage 5P4 left SEMAN works only in early 1990.

15 This was fitted in early 1988 to Boeing 707 number FAP 319 from Escuadrón de Transporte 842.

Lima, summer 1980. A pair of Mirage 5P3 at the tarmac of Grupo Aéreo N° 8 in Callao. (Iraola)

The purchase of additional deltas along with an effective maintenance program allowed the FAP to activate a second Mirage 5P Squadron. EA 612 operated alongside Sukhoi Su-22 Fitter F fighter-bombers from La Joya airbase.

This modernization program gave the Peruvian Mirage 5s the capability to carry Matra Magic heat-seeking air-to-air missiles. The inertial platform provides self-contained navigation, eliminating the need for external radio aids such as ADF while the laser increases delivery accuracy on low-level penetration missions. Most significantly, the Navattack update introduced automatic weapon release, with several modes available for use with guns, rockets, low-drag and retarded bombs. Waypoints and targets are pre-programmed into the system's memory allowing low-level, high-speed, single-pass attacks on designated targets.

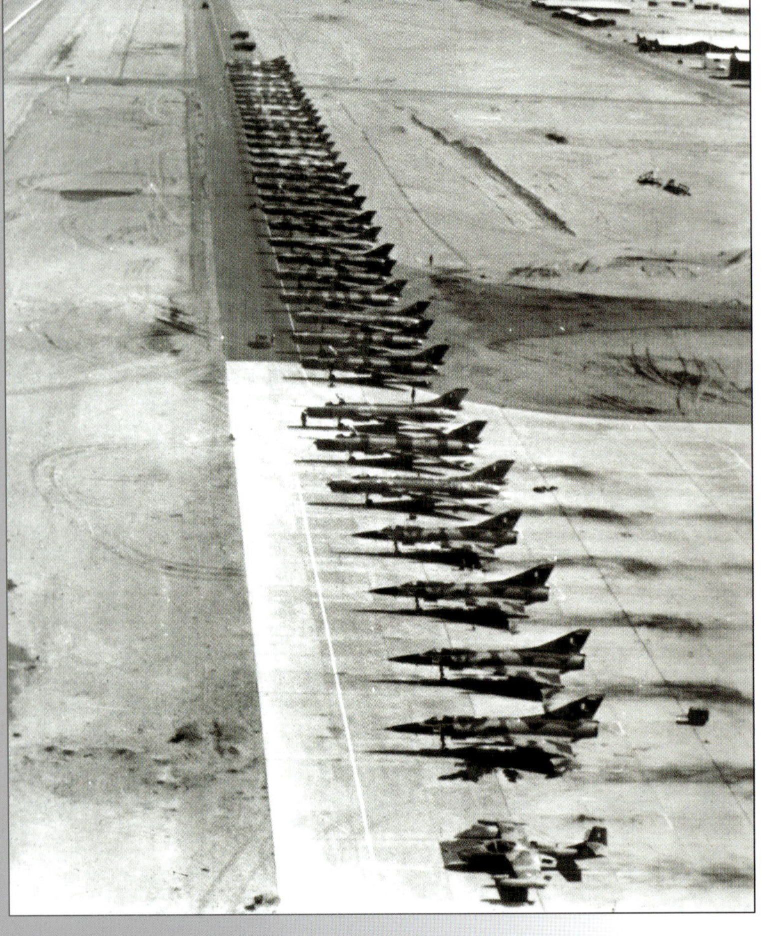

The number of Mirage 5 lost by the FAP increased during the second half of the eighties. On March 10, 1987 Mirage number FAP 189 was lost during a practice flight, its pilot being able to eject safely. However, during the month of May of the next year the EA 611 suffered a major blow with two incidents taking the lives of valuable officers. On May 18 Mirage 5DP4 number FAP 196 crashed due to an engine problem, causing the death of one of its occupants and, a few days later, Mirage 5P4 number FAP 114 crashed against El Fronton Island on the coast of Callao due to bad weather, killing Capt. Rebaza, its pilot.

Operations in the "Alto Cénepa"

During the beginning of the nineties, Ecuador and Peru once again became involved in border disputes, this time over a still unmarked 75-mile area in the so-called "Cordillera del Condor" mountain range located between the Cénepa and Santiago rivers. On January, 1995, the disputes developed into an armed conflict when the Peruvian Army discovered that not

This impressive aerial view from La Joya AB shows ten Mirage 5P and 13 Su-22A, NATO code name Fitter F. (De la Puente)

3rd year Cadet Ricardo Iraola Villa walks in front of a lineup of Mirage at Las Palmas airbase in January 1982. A few months later these machines would be flown to La Joya airbase, where they received FAA insignias and then were flown to Tandil airbase, Argentina. (Ricardo Iraola)

Lineup of Mirage 5P3 from EA 612 seen parked at "Capitán Renán Elías Olivera" AB in Pisco during a navigation exercise held in the early eighties. (DIRIN)

Profile view of a Mirage 5P3 at Pisco.

only had Ecuadorian Army elements infiltrated Peruvian territory, but established a number of false border vigilance posts as well.

The global economy crisis of the eighties echoed hard in Peruvian domestic economy, barely recovering from 12 years of a socialist dictatorship and its inherently poor economic management. Unfortunately, these effects were magnified by the internal convulsion generated by the "Sendero Luminoso" (Shining Path) terrorist group as well as the corruption and disastrous economic measures during President Alan Garcia Perez's administration. Additionally, an allegedly irregular transfer of large stock of spare part to Venezuela, whose Air Force was by then immersed in a Mirage system upgrade process, took place which directly affected the EA 611 operations causing a marked decrease in operational readiness in all units on account of low maintenance as well as a reduction in flight hours for aircrews. This was aggravated by the severe budget restrictions and cuts caused by the counterinsurgency (COIN) operations, which meant that, by the early nineties, the Peruvian Air Force suffered an overall reduction on its operational capabilities.

Months earlier, on October 10, 1994, a M5DP4 was lost due to an engine malfunction and, after investigations, FAP High Command grounded the entire Mirage fleet until every single airframe was inspected, thus paralyzing flight activities of EA 611. The investigation and inspection of all airframes was completed by December of 1994, but flights did not recommence immediately, as the required factory certification paperwork did not arrive due to problems with pending payments owed by the Peruvian government to Dassault Aviation[16] from the purchase of twelve Mirage 2000P fighters in 1984.

With an operational but not "certified" fleet of five Mirage 5Ps (four Mirage 5P4 single seat and one Mirage 5DP3 two-seater), put back in the air with great effort by the ground crews by cannibalizing other aircraft, EA 611 launched a "refreshment" program with a minimum of aircrews in order to bring them to "combat ready" status. The FAP situation at the time was clearly different from 1981, but this did not prevent the Air Force to accomplish its assignments. This time, the Sukhoi Su-22M3 Fitter J from EA 111 and the A-37Bs from EA 711 were called to carry the burden of the CAS and strike missions in support of the ground offensive, while EA 611 was kept in reserve at Chiclayo on stand-by alert, ready to perform deep strike missions against Ecuadorian targets in the Guayaquil bay area if the localized conflict escalated into all-out war. Nevertheless, television reports showed several images of Mirage 5P4 aircraft fully armed either with Matra JL-100 68mm SNEB rocket launcher/fuel tank combinations or with RPK-10 bomb rack/fuel tanks containers, as well as Vympel R-3S air-to-air missiles, sitting on the tarmac with pilots strapped to their seats in

16 Company name from 1990 onwards.

their cockpits. After a series of victories and defeats for both sides a cease fire was agreed with the mediation of neighboring countries and the United States of America, opening the door for peace talks which would span for almost three years. In the meantime the FAP, still licking its wounds from the losses sustained during the short but violent conflict, launched a capability-recovery program which included the recovery and planned upgrade of the Mirage 5 fleet in order to face an eventual renewal of hostilities against Ecuador in the event of failed peace negotiations. The aftermath of the conflict brought, however, a new loss to the Mirage 5P fleet when a Mirage 5P4 crashed in Chiclayo during a training flight, killing its pilot.

Between the years 1997/98 and as part of *Operacion Grajo*, the Mirage fleet was subjected to an exhaustive airframe check, engine revision and a passive defense systems upgrade in the form of the integration of IMI SAMP chaff-flare dispensers and SPS-20 radar warning receivers. Finally, the whole fleet received a new paint scheme of green-grey colors. In the meantime, EA 611 carried out numerous training sorties against simulated Ecuadorian strategic targets at Reque gunnery range, in the event of a renewal of hostilities. However, the arrival of 18 MiG-29S fighter-bombers and their Beyond Visual Range (BVR) capabilities in January 1997[17] apparently put enough pressure on the Ecuadorean authorities to convince them to sign the peace agreement at Brasilia later that very same month.

The end of the road

With the arrival of the MiG-29 fighters and the subsequent re-activation of EA 612, the Mirages from EA 611 were reassigned to La Joya AB airbase in Arequipa, from where they operated sporadically. Between late 1997 and early 1998, Dassault Aviation approached the Peruvian Government offering to improve the Mirage 5P4 strike capability with an upgrade which included the fitting of a Thales 3000 laser-gyro Inertial Reference System, a Thomson-CSF ATLIS II Laser Designation Pod and associated cockpit equipment, landing gear reinforcement as well as a new wheel braking system. Unfortunately, lack of funds prevented any further interest in this upgrade. Shortly afterwards, a couple of *Dardos* were employed to test bombs fitted with *Lizzard/Opher* precision guidance systems, part of the *Whizzard* family of precision guided munitions developed by ELBIT of Israel.

17 Purchased from Belarus that previous December, these were second-hand machines in very good condition and with a lot of flying hours remaining on their engines and airframes.

Formation of Mirage 5P3 from EA 612 during a navigation exercise realized as part of the PDE 1980. (DIRIN)

These kits could be installed in conventional Mk 80, Mk 82, Mk 83 and Mk 84 "dumb" bombs, and the combination of infra-red (*Opher*) and laser (*Lizzard*) guidance systems provided the FAP with point-kill targeting capabilities against moving targets such as vehicles, as well as hardened objectives. However, as neither the Mirage 5P4s nor the Mirage 2000P fleet had the associated electronic devices required for their employment, guidance and tracking had to be done by means of a second aircraft fitted with these capabilities in the case of the *Opher* and/or by laser air or ground designation for the *Lizzard*.

Continuous budget restrictions continued to harm FAP operational capabilities during the new millennium, forcing its High Command to withdraw several types from service that, under different circumstances, would have been useful for longer periods of time, and such was the case of the Mirage 5P fleet. During 2001/2002 the remaining Mirage 5P4/DP4 airframes were subjected to long-term storage at Las Palmas hangars, while awaiting their possible sale to any potential customer. No interest was shown and, as of July 2011, the thirteen remaining units could still be found at the same location.

Mirage 5P Fleet in numbers

During the span of more than 30 years the Peruvian Air Force operated a total of 40 Mirage 5P – 34 single seat and six two seat trainers – acquired under the terms of ten contracts, code-named *Martillo* (Hammer) I to X, detailed as follows:

- *Martillo* I: signed on July 20, 1967, for ten Mirage 5P (N° 182 to 191) delivered between May 1968 and August 1967, plus two Mirage 5DP (numbers FAP 196 and FAP 197) delivered in October and November 1968, respectively;
- *Martillo* II: signed on January 30, 1968, comprising four Mirage 5P (numbers FAP 192 to FAP 195) delivered in pairs between October and November 1969;
- *Martillo* III: signed on November 31, 1973 for the purchase of a single Mirage 5DP (number FAP 197 – 2[18]). Delivered in February 1974;
- *Martillo* IV: signed on July 30, 1973, for the procurement of seven Mirage 5P (correlative numbers FAP 101 to FAP 107) which were delivered between October 1975 and January 1976, plus a single Mirage 5DP (FAP 198) delivered in October 1975;
- *Martillo* V: signed on November 30, 1973, for the procurement of a single Mirage 5P (FAP 188 – 2). Delivered in March 1976;

18 Replacement aircraft.

A Mirage 5P3 flying at high speed over typical landscape found on the Peruvian north-eastern territories during the type participation in the 1981 conflict. (Arraez)

- *Martillo* VI: signed on May 15, 1974, for the purchase of seven Mirage 5P3 (correlative numbers FAP 108 to FAP 114) delivered between August 1976 and February 1977 plus a single Mirage 5DP (number FAP 199) delivered in February 1976;
- *Martillo* VII: signed on December 30, 1977, for the purchase of a single Mirage 5P3 (FAP 193 – 2). Delivered on May, 1977;
- *Martillo* VIII: signed on April 30, 1976, for the purchase of a single Mirage 5P3 (FAP 184 – 2). Delivered in September 1978;
- *Martillo* IX: signed on February 28, 1977, for the purchase of a single Mirage 5P3 (FAP 191 – 2). Delivered in October 1978;
- *Martillo* X: signed on June 8, 1981, for the procurement of a pair of Mirage 5P4, FAP 190 (2) and FAP 194(2) as well as a single Mirage 5DP4, FAP 198(2) as attrition replacements, plus 15 conversion kits to upgrade a similar number of FAP Mirages to the P4 standard, a process completed between the years 1985 and 1990.

An impressive line of aircraft from EA 611 and EA 612 fills the tarmac of "Capitán José Abelardo Quiñones Gonzales" *AB in Chiclayo during the participation of these units in the so called* "Falso Paquisha" *conflict. Of interest are the R-3S missiles fitted to some aircraft as well as the two airframes still in original AdA tactical grey/green scheme in the extreme left. (DIRIN)*

An impressive array of FAP attack aircraft parked at Grupo Aéreo N° 8 *tarmac in early 1982. At least 13 Mirage 5P3/DP3 from EA 611 can be observed in this picture.*

The upper surfaces paint scheme of the Mirage 5P3 is seen in advantage on this picture taken from the control tower at Chiclayo, during 1982. (Arraez)

FAA gave good use of the ten Mirage 5P purchased to Peru in 1982. This Mirage 5 "Mara" shown some interesting features such as the presence of radar warning receivers in front and rear fuselage, as well as the use of supersonic fuel tanks still in original FAP desert paint scheme. (Rivas)

The Peruvian Mirage IIICJ, complete with desert camouflage and fake wing insignia and four digits FAP number, is shown while aboard the freighter that carried it and 22 similar aircraft, from Israel to Argentina in 1982. (Rivas)

Left and below: *The P4 upgrade provided the FAP with an aircraft capable of deep strike capabilities. A Mirage 5P4 seen during trials at the SEMAN in Las Palmas after completing its upgrade as part of* Proyecto Pegaso. *(DIRAF)*

Beautiful shot of a Mirage 5P4 identified with FAP 194 equipped with 1700 liter auxiliary fuel tanks flying over the Andes mountain range. (Mejia)

With the arrival of Boeing 707 number FAP 319, which was specially modified with a French fuel tanker conversion kit in the late eighties, the FAP began aerial refueling training. Up to 3 aircraft could be refueled in the air at any given time. Here a Mirage 5P4 with no underwing stores being refueled. (Cruz)

A Mirage 5P4 armed in air to air configuration provides escort to a FAP Boeing DC-8 from Escuadron de Transportes 842 carrying his holiness Juan Pablo II to Lima in February 1985. (FAP)

Pilots trained in refueling approach and contact procedures aboard the Mirage 5DP4 before progressing to "solo" attempts. Here a pilot flawlessly approaches its probe-less M5DP4 to the refueling hose while a single seat Mirage 5P4 pilot awaits its turn in the background. (Cruz)

Mirages from EA 611 seen parked on the tarmac of "Capitán José Abelardo Quiñones" AB in Chiclayo during the late eighties. (DIRIN)

Mirage 5DP4 number FAP 198 and Mirage 5P4 number FAP 187 await the beginning of a new training mission in Chiclayo during 1986. (Arraez)

Mirage 5P4 trio in CAP configuration armed with Vympel R-13M AA missiles. Note that the last aircraft from the formation doesn't have the IFRP fitted. (Mejia)

Spectacular shot of a Mirage 5P4 landing at Chiclayo. (EA 612)

Beautiful sunset view of a four ship Mirage 5P4 formation. (EA 612)

Interesting picture of a fully loaded Mirage 5P4 FAP number FAP 110 showing the RPK-10 fuel tank/bomb rack combination to advantage. This aircraft was the pattern for the Proyecto Pegaso Mirage 5P4 upgrade. (Mejia)

Mirage 5P4 number FAP 101 was modified with the installation of the Thomson-CSF SHERLOC digital RWR as well as chaff/flare containers during the year 1989. Unfortunately the project did not proceeded beyond the evaluation stage due to lack of funds. (Mejia)

Mirage 5P4, FAP 192 armed with eight Mk 82 bombs and two R-13M Air-to-Air missiles. Las Palmas, 1994. (Alcorta)

Another view from the same aircraft. (Alcorta)

Las Palmas airbase, 1994. A Mirage 5P4 towed by a "mule" pass in front of a Mirage 2000P resting at the tarmac of this base. (Alcorta)

By the end of the nineties only a handful "Dardos" remained in airworthy condition. Four aircraft perform a low level passage over La Joya airbase in 1998. (Mejia)

As preparative measure for a potential renewal of hostilities between Peru and Ecuador some of the Mirage 5P4 were given a new paint scheme, more suitable for operations over the jungle than the original sand/dark earth scheme. One of such aircraft prepares to land at Las Palmas. (Moeri)

Three old warriors approaching to the end of their careers await yet another mission before the end. A BAC Canberra B.12 from EA 921 and a pair of Mirage 5P4 from EA 611 seen on the tarmac of Las Palmas in the late nineties. (Cruz)

During PDE 2000 FAP Mirage 5DP4 took place on the evaluation of the Elbit's Wizzard *precision munitions delivery system, with the guidance made by ground designators. Here, a Mirage 5DP4 from EA 611 departs the runway at "Renan Elias Olivera" AB in Pisco for a weapon release exercise.*

Near the end of its operative life, the Mirage Mirage 5P4s received a black paint cover over their nosecones, in an attempt to confuse any potential adversaries with the more advanced Mirage 2000P. Las Palmas year 2001. (Alcorta)

Several views from Mirage 5DP4 FAP 198, preserved at Las Palmas airbase. (Herrera)

Mirage 5P, FAP 183, Escuadrón de Caza 611, *Chiclayo airbase in 1968. Camouflage:* Gris Bleu Fonce *(Medium Grey) and* Gris Vert Fonce *(European Green) upper surfaces with* Gris Bleu Moyen Clair *(Light Blue Grey) undersurfaces. This was the second airframe of this type to arrive in Peru in 1968 serving with the FAP until June 1982 when it was transferred, together with other nine airframes, to Argentina.*

Mirage 5P3, FAP 191, Escuadrón de Aviación 611, *Chiclayo airbase in 1981. Camouflage: Pale Stone and Dark Earth over Light Blue Grey. Reduced size numbers painted in black below national flag on vertical surfaces. The aircraft performed CAP and fighter sweep missions over the Condor Mountain Range fitted with a pair of Vympel R-3S* Atoll *Air-to-Air missiles.*

Mirage 5P4, FAP 110, Escuadrón de Aviación 611, *Chiclayo airbase in the early nineties. Camouflage: Pale Stone and Dark Earth over Light Blue Grey. Black nose radome. Reduced size numbers painted in black below national flag on vertical surfaces. The aircraft is seen fitted with a pair of JL-100 rocket launcher/external fuel tank pods as well as an identical number of Vympel R-3S* Atoll *Air-to-Air missiles for self-defense.*

Mirage 5P4, FAP 192, Escuadrón de Aviación 611, *Chiclayo in 1996. Camouflage: Dark Grey, dark Green and Medium Green upper surfaces over Light grey undersurfaces. Beginning in the summer 1995 at least half a dozen Mirage 5P4 airframes were put back in airworthy condition in anticipation of a renewal of hostilities with most of them receiving a new, more appropriate, paint scheme, commonly identified as the "forest scheme".*

Mirage 5DP4, number FAP 198, Escuadrón de Aviación 611, *Chiclayo, early nineties. Camouflage: Pale Stone and Dark Earth over Light Blue Grey. Black nose radome. This aircraft was purchased from Dassault-BA in June 1981; while another Mirage 5DP (FAP 197-2) was later upgraded to the 4 standard as part of the* Pegaso *project. While fully weapon capable, the pair of twin seaters were used only for training purposes.*

MIRAGE 2000P/DP

The purchase of Dassault-Mirage F.1 and Northrop F-5 Tiger II fighters by the air forces from Ecuador and Chile, respectively, evidenced the need to upgrade the Peruvian Air Force fighter force with a latest-generation modern air-superiority aircraft. In January 1979 the FAP High Command, in coordination with the *Consejo Nacional de Defensa* (National Defense Council), issued the Planning Directive 17-1 aimed to replace obsolescent aircraft just withdrawn from front line service, such as the F-86F "Sabre" and Hawker F.52 "Hunter" and authorizing the launch, in early 1981, of a re-equipment program aimed to select the new FAP front line fighter for the next two decades. Therefore, in the spring of 1981 the FAP HQ ordered the creation of a technical committee tasked with the evaluation process of the technical characteristics, costs and availability of the several proposals submitted by manufacturers worldwide, amongst them Israel Aircraft Industries with its IAI Kfir C-7, AMD-BA offering the Mirage 2000, General Dynamics with the F-16 "Fighting Falcon" and, finally, Mikoyan-Gurevich with the MiG-23 "Flogger". After completing the evaluation process the commission found the Dassault-Breguet Aviation bid as the winner, due to the outstanding performance and capabilities of the Mirage 2000 over the remaining contestants. Therefore, on October 11, 1982, Peruvian government signed the Jupiter I & II contracts between his government and the French SNECMA, Thomson-CSF and AMB-BA aviation firms, for the procurement of 26 Mirage 2000 fighters – 22 single seat and 4 two-seater airframes – for a total of 4,564,000.00 francs.

With machines already under construction, FAP sent a number of its most seasoned pilots to Mont-de-Marsan AB, home of the AdA 118 Experimental Centre. Maj. Felipe Conde Garay and the captains Donovan Bartolini, Ricardo Vilchez and Guido Zavalaga arrived in that location on May 1985 and followed, during the next six months, a very intensive theoretical-practical training program. Meanwhile, the first machines rolled out of the factory and on May 30, 1985, Mirage 2000DP number 193 was symbolically handled over to Lt. Gen. Cesar Enrico, FAP Chief in Command, and Lt. Gen. Percy Velarde, head of the *Direccion de Material de Guerra*[19] on a ceremony

19 Military Equipment Bureau.

A nice in-flight shot of Mirage 2000DP number FAP 195 taken over France during tests prior to its delivery in mid 1985. Note the absence of Peruvian national roundels and flags. (DIRIN)

Several shots depicting the shipping process of the first four Mirage 2000P to Peru, two Mirage 2000P single seat fighter identified with numbers FAP 050 and 054 and two Mirage 2000DP two seat trainers identified with numbers 193 and 195. These aircraft left France in late October 1986 aboard French freighter François Billon. *(DIRIN)*

held during the Le Bourget Air & Space Exhibition. Shortly after, on June 7, *Jupiter* III contract, which covered support equipment as well as a Dassault Falcon 20F corporate jet in VIP configuration, was subscribed, raising the original obligation from 4,564,000.00 to 4,960,000.00 francs – roughly 650 million dollars at the time. A further contract, identified as *Jupiter IV*, was expected to be signed afterwards and would cover weaponry, support equipment and a flying simulator.

Things would take an abrupt turn for the Mirage 2000 deal when, on his very first speech after assuming the office on July 28, 1985, the President announced the reduction of the Mirage order from the original number of 26 to just 12 machines, and cancelling the option for the ten additional airframes, a decision that caught the Air Force High Command officials by surprise, as there was no previous word from the president about this. In fact, President Garcia had taken this decision one month earlier, again without consulting the Air Force. This controversial decision was the

starting point for a sound – and still unsolved to this date – corruption scandal involving shady business between politicians, diplomats, arm dealers, bankers, etc. details which, unfortunately, go beyond the scope of the present work.

While politicians took care of the *Jupiter* contract renegotiations, FAP officers already deployed in France began their training on their new mounts. On October 8, 1985, Maj. Felipe Conde and Capt. Ricardo Vilchez became the first qualified FAP Mirage 2000 pilots but shortly after completing their first flights these men were ordered to return to their country since the AMD-BA refused to ship the airframes until the end of negotiations with the Peruvian government. Finally, with the signature of the much re-arranged Jupiter IV contract in mid-1986, the Peruvian air force agreed to take delivery of just ten Mirage 2000P and two 2000DP fighter-bombers along a single Falcon 20F corporate jet while the flight simulator, support equipment and advanced weapons such as infra red and radar guided missiles were left out of the deal. The lack of this critical equipment would cost the FAP dearly a decade later, during the 1995 conflict between Peru and Ecuador.

On November 20, 1986, the first four airframes arrived to Pisco harbor aboard the French freighter *François Billon*. A few days later, the semi-assembled airframes were taken by ground to the nearby "Capitan Renan Elías Olivera" AB in Pisco. With the delivery of these first airframes the recently activated EA 412 "Halcones" began an intensive training program and, by the time the last two airframes were delivered in mid-1988, the unit had 18 fully qualified pilots.

El Alto Cenepa

The Mirage 2000P saw its baptism of fire during the January 1995 conflict between Peru and Ecuador in the course of which the deltas from EA 412 carried out no less than 80 sorties for a total of 110 flight hours. The missions assigned to this unit were to provide CAP over the operations area as well as over sensitive positions such as the oil refinery at Talara and the nearby airbase, headquarters of GA N° 11. The unit accomplished the orders in the mist of adverse weather conditions and, more importantly, operational constraints caused by the budget restrictions which prevented EA 412 to adequately complete the aircraft maintenance programs according to schedule, as well as preventing pilots from reaching their PDE´s minimum flight hours. As consequence the unit could only muster a total of seven airworthy machines during the conflict, distributed between Chiclayo and Talara. In addition to these shortcomings the EA 412 had to deal with a number of

Le Bourget, May 30, 1985. FAP Mirage 2000DP number FAP 193 was exhibited during the French Air and Space Saloon prior to its delivery. (Peijmen)

communications, radar and electronic systems malfunctions which effectively hampered the effectiveness of the aircraft and the use of its Matra R550 "Magic 2" missiles, purchased in late 1993. These limitations became the chief culprits for at least three of the losses sustained by the FAP. On February 10, the FAE claimed two FAP Su-22s shot down by Matra R550 "Magic 2" missiles launched by marauding Mirage F.1JE fighters which ambushed the Peruvian attack aircraft as they approached their target. A patrol of Mirages from EA 412 was deployed to the area ahead of the Sukhoi but it was too late for it to react after the presence of the intruders was revealed and these had performed their attack. While the Mirage 2000Ps managed to track the FAE aircraft with their radars, the intruders maneuvered low and at full speed to return to the safety of their own lines.

With the cease fire and the beginning of peace talks between Ecuador and Peru, FAP High Command sought for an urgent increase of capabilities of its fighter arm. Since a Mirage 2000 upgrade would be a slow process and there was a desperate need for fighters readily available, the

A pair of Mirage 2000P from Peruvian Air Force follows the lead of a FACh Mirage 50 Pantera which welcomed the Peruvian aircraft as they arrived to Santiago de Chile to take part on the FIDAE 1994 Air and Space Fair held on the Chilean capital. (Rivas)

In the cockpit of FAP 053 Maj. Mosca performs the pre-flight procedures for its acclaimed aerial display at FIDAE 1994, Santiago de Chile. (Alcorta)

Peruvian Government sought, as a matter of urgency, to acquire fighters overseas. The short process ended with the purchase of 18 Mikoyan Gurevich MiG-29S *Izdelie 9.13S* from Belarus. In parallel, FAP began negotiations with Dassault Aviation looking to upgrade its Mirages, but the excessive cost and the refusal by Dassault Aviation to perform any further work on them until Peru honored the outstanding debts with the company prevented further action.

Operations with *Escuadron Aereo* 412 1996-2013

Training of EA 412 pilots begins with the *Curso de Entrenamiento Operativo*, which is followed by the PDE, allowing each pilot to reach their required annual flying hours. Additionally, to help in the training and formation of aircrews, a flight simulator – manufactured by Sogitec and bought in late 1990 – allows crews to train in flight procedures such as navigation and emergencies as well as combat missions. This flight simulator was modernized around the year 2006, and operates inside an 8m sphere containing the screen, while an environmental system presented ground, sky, horizon, acceleration, clouds and rain, as well as sound effects for the environment.

Since the arrival of the MiG-29S/SE 'Fulcrum C' fighters in January 1997, EA 412 have taken part in several joint exercises with EA 612 pilots, performing Dissimilar Air Combat (DACT) procedures which included confrontation between BVR vs. non BVR-capable machines. Also, during 2002 and 2007 pilots from EA 412 took part in several international training operations, such as UNITAS 45-04, PASSEX 2004 and *HALCON-CONDOR 2007* where they operated alongside US Navy F/A-18E/F Super Hornets from the 11 Carrier Air Wing, based on US aircraft carrier USS *Abraham Lincoln* (CVN-72), obtaining valuable experience in modern air combat tactics. Additionally, during HALCON-CONDOR exercises performed in February 2007, Mirage and MiG pilots were able to face, for the first time ever, their nemesis, materialized in the form of five F-16C/D Block 40 from the 4th and 34th Fighter Squadrons, 388 Air Wing based at Hill Air Force Base, in Utah. Despite the fact that official results from the drill remain classified, unofficial reports state that the Peruvian MiG-Mirage duo scored some "kills" against their more capable "foes";

Mirage 2000P number FAP 053 leaves the runway at "Los Cerrillos" AB with Cmdr. Mosca at the controls. It was the first Mirage 2000 presentation in South American skies. (Alcorta)

most importantly, Peruvian pilots were able to obtain invaluable experience by testing their machines and themselves against late-block Falcons, similar in capabilities to these equipping the *Fuerza Aérea de Chile* (FACh), which at the time is the FAP´s most likely threat.

Present and future

In the year 2003 the FACh introduced the Lockheed Martin (formerly General Dynamics) F-16C/D Block 50 into service, a move that put considerable pressure into the FAP high command as the Fighting Falcon clearly outmatched the Mirage 2000P thanks to its advanced avionics and equipment, not to mention weaponry. As response to the new threat the air force high command began the studies aimed to upgrade the FAP Mirage fleet to the Dash 5 version, an investment of at least 300 million US dollars at the time. Politicians, however, did not share the military's point of view and concerns, denying the funds for the much needed upgrade until late 2009 when, finally, the Peruvian Government reached an agreement with Dassault Aviation, for fleet inspection, required to keep the fleet in operational and certified condition as well as to perform any future upgrades. The agreement opened the door for the VP 5[20] inspection, at a cost of 140 million US dollars, followed shortly after by the first of three-level upgrades. While the VP 5 included the

20 *Visite Periodique* or Grand Visit 5.

FAP 053 taxies back to its parking spot after completing a widely acclaimed display under the command of the experienced Peruvian pilot. (Alcorta)

Las Palmas airbase, July 1994. Mirage 2000P FAP 050 armed with practice bombs seen during a display for the Peruvian military aviation day. (Alcorta)

replacement of electrical wiring, weapon hard points, inspection of ejection seat systems as well as outdated integrated circuits, the first limited upgrade included a complete engine revision, the replacement of outdated SAGEM UNI 52 INS with a Thales Totem 3000 gyro-laser inertial reference system, of improved performance. Also, the landing gear bay has been modified as well as the landing gear, which received an SPAD braking system. A new paint scheme, consisting of a dark gray overall finish, similar to that employed by the Dassault *Rafale*, had been applied to the aircraft as they completed their upgrade.

Weapons and equipment

Although still relegated to deep strike role due to the obsolescence of its RDM radar, for CAP sorties FAP Mirage 2000P carry the Matra R550 Magic II infrared air-to-air missile in addition to the standard built-in 30mm DEFA 554 30mm cannons; for air-to-ground operations, however, the situation differs substantially, as the Mirage incorporates a deadly plethora of weapons, including the Mk.82 *Snakeye* and 250 kg SAMP 25FI general purpose bombs, BME-330[21] cluster bombs and either DURANDAL or BAP 100 anti-runway bombs. During 1997 a limited weapons upgrade was negotiated with France and Israel, including the Thales ATLIS II Laser Designation Pod, but lack

21 Manufactured by Expal in Spain.

Arequipa, summer 2003. FAP Cmdr. Chavez Cateriano performs a gear down approach for the cameraman over the Peruvian southern desert. (Chavez Cateriano)

Impressive lineup of FAP aircraft at La Joya airbase. One Mirage 2000DP, nine 2000P and five Sukhoi Su-22M get attention from ground crews before operations. (Moeri)

With the arrival of the BVR capable MiG-29S air superiority fighter in 1997 the crews from EA 412 and EA 612 programmed annual dissimilar air combat situation exercises to improve their capabilities. (EA 612)

of funds ultimately put an end to negotiations with Thales. Eventually ELBIT´s *Whizzard Opher/ Lizard* IIR or Laser guided bombs were purchased and integrated as part of the Mirage 2000P weapons stock, giving the aircraft pinpoint strike capabilities, albeit with the aid of an external aerial or a ground designator. Meanwhile, during the year 1998 and as direct result of FAP experiences during the 1995 conflict, Thomson-CSF TMV-004 *Cayman* Electronic Counter Measures pods were acquired as well as the MBDA *Spirale* Decoy and Electronic Counter Measures System, offering an improved level of protection against air and ground based threats.

Mirage 2000P, FAP 053, La Joya airbase in the late nineties. Camouflage: Sand and Chocolate Brown upper surfaces over Light Blue Grey undersurfaces. Black radome. In the late nineties this aircraft was subject to a ECM suite upgrade which included the fitting of the Spirale Chaff/Flare dispenser system, located on the wing root training edges.

Mirage 2000P, FAP 050, La Joya airbase in 2012. Camouflage: Dark Gull Grey overall scheme with Dark Grey national markings and numbers. This scheme was introduced in late 2009 as the FAP Mirage 2000P fleet was subject to a major inspection.

With the arrival of the BVR capable MiG-29S air superiority fighter in 1997 the crews from EA 412 and EA 612 programmed annual dissimilar air combat situation exercises to improve their capabilities. (EA 612)

A view of a Mirage 2000DP and a Mirage 2000P from EA 412 being escorted by an F/A-18C from the 94 "Mighty Shrikes" Strike Fighter Squadron (VFA-94) during the PASSEX 2004 exercises. (Chavez Cateriano)

FAP 054 seen from below. Note the presence of the Spirale ECM system fitted to the FAP Mirage 2000 fleet during the late nineties. (Chavez)

Night operations are part of EA 412 continuous training. A Mirage 2000P identified with number 062 awaits the beginning of another training mission.

After the 1995 Peru – Ecuador conflict, FAP Mirage 2000P fleet received a limited ECM and weapons upgrade, which included SPIRAL electronic counter measures system as well as the Wizzard weapon system, manufactured by ELBIT in Israel. (Alcorta Stoll)

By the end of 2003 the FAP Mirage 2000 fleet received a new paint finish similar to that employed by the French AdA. Here, Mirage number FAP 193 can be seen during an "open doors" held in Las Palmas shortly after the culmination of the "Halcon-Condor" exercises, in February 2004. (Herrera)

The clear skies of the Andes appear to blend with the Mirage 2000P colours. The aircraft, identified as FAP 063, was part of a FAP display during at an airshow organized at "Teniente Alfredo Rodriguez Ballon" international airport, Arequipa, 2007. (Mongrut)

"Teniente Alfredo Rodriguez Ballon" international airport, Arequipa, May 2011. As part of the 25th anniversary of the Mirage 2000P in EA 412 service, a single airframe was decorated with a neat "Hawk" illustration on its tail.

Mirage 2000P number FAP 064 banking over the typical landscape found on Peruvian southern territories. The Spiral chaff-flare dispensers, retrofitted to the aircraft in 1998, can be seen in advantage. (Carlos Chavez Cateriano)

La Joya AB, early 2011. The pilot of Mirage FAP 060 from EA 412 prepares to land after completing a live air to air missile firing drill. (Carlos Chavez Cateriano)

Beginning in late 2010 the Mirages from EA 412 began to receive a coat of light gray on their radomes, similar to those employed by other Mirage 2000 users. A pair of EA 412 aircraft took part in the air show held in Tarapoto international airport in late 2011. (Yui)

A formation of "Halcones" from EA 412. A trio of Mirage 2000P from EA 412 and pair of Aermacchi MB.339 advanced trainers from EA 513 seen during training at La Joya, in Arequipa, in late 2011. (Chavez Cateriano)

Starting in late 2009 the Mirage 2000P/DP fleet was subject to an airframe inspection and life expansion program. The airframes received a new low-visibility paint finish after completing their inspection.

Lineup of Mirages from EA 412 at Las Palmas in July 22, 2013.

CHAPTER FOUR
ALOUETTES AND FALCONS

SUD-AVIATION SE.3130/SA.318C ALOUETTE II

During the late fifties Peruvian Air Force requirements for a scout, rescue and light transport helicopter led to the purchase of five SE.3130 Alouette II aircraft from France in late 1959. The aircraft were delivered by early 1960 and upon arrival they were immediately assigned to the 71 *Escuadrón de Rescate*, based at Limatambo. Only a couple of pre-1960 numbers are known for these machines, 556 and 574, but after the year 1960 reorganization the aircraft became identified with correlative numbers FAP 60-601[1] to 60-605[2], as well as FAP 60-610.

Insufficient training took a heavy toll on the Alouettes. On May 30, 1962, FAP 604 crashed during a training flight, killing its two occupants. Better luck had the crew from FAP 603 which crashed seventeen days later near Lima after an engine failure. The pilot managed to keep control of the ship during descent and, miraculously, he and his copilot were only slightly injured in the ensuing crash.

In late 1963 the 71 ER was redeployed to the newly activated air force base located in the "Jorge Chávez" international airport, Callao, where it became *Escuadrón Aéreo* 332. On May 20 of 1964 Alouette II number FAP 602[3] crashed due to a pilot mistake, wounding its two occupants. A few months later, on August 2, a mechanical failure caused the loss of SE.3130 number 605 a few miles east of Caraz, killing Alférez FAP José Cavero and 3rd class S.O. Juan Serrano Vásquez.

1 The prefix number was employed during the first years of the new registry number use and it indicated the airframe year of arrival. Its use was discontinued by 1964.

2 Co/Nu 1486/236C.

3 Co/Nu 1438/204C.

A single SE.3130 Alouette II scout and liaison helicopter parked in the tarmac of Grupo Aéreo *N° 8 at Callao in 1971.*

By early 1966 only a couple airframes remained airworthy due to the several accidents sustained by the fleet. In February 20, 1967, FAP 610[4] was also written off in an accident, leaving FAP 601[5] as the only flyable machine. This helicopter survived in airworthy condition until the early nineties when it was sold to a Japanese citizen[6].

"Alouette II" in service with *Ejército del Perú*

In early 1974 the Peruvian EP ordered 9 Aerospatiale SA.318C.1, which were delivered during the second half of the year. These aircraft received in tree batches receiving the numbers EP N° 312, – 315, – 318; – 321, – 324, – 327 and – 334, – 337, – 341. The aircraft were assigned to the newly activated *Batallón de Reconocimiento y Ataque 811.*

On March 1, 1976, Alouette II EP N° 341[7] suffered a mechanical failure and crashed in the mountains near Huaraz, with the loss of all aboard. Despite this incident the EP crews appreciated the aircraft positive flight characteristics and good handling.

Tear and wear began to affect the fleet and on February 14, 1990, an engine trouble caused the loss of Alouette N° 315[8] and its two man crew during a training flight in a beach in Ventanilla, located North West of "Jorge Chávez Dartnell" international airport in Callao.

The EP Alouette II fleet gave a good account of itself and continued to serve until the early twentieth century, with the last airframe being withdrawn from service in the year 2004.

SUD SE.3160/SA.316C "Alouette III"

In July 1961 Peruvian air force ordered from the French SUD Aviation Company twelve SE.3160 Alouette III utility and light attack helicopters to equip its newly formed 71 *Escuadrón de Rescate* (71 Rescue Squadron – 71 ER). SUD Aviation delivered four of these machines in January 1962, followed by the remaining eight in May of that year.

4 Co/Nu 1074.
5 Co/Un 1421/194C.
6 Sold as JA-9826.
7 Co/Nu 2362.
8 Co/Nu 2345, French civil registry F-WKQD.

A nice view of OB-E-28, an SE.3160 Alouette III helicopter belonging to Escuadrón Aéreo *332 seen parked at* Grupo Aéreo N° 8 *in Lima. The aircraft is finished in orange with grey undersides.*

The Alouette III could be armed with rockets, as evidenced in this picture, to perform light support in behalf of infantry operations. (DIRAF)

A nice color picture showing an aerial display by a trio of Alouettes during the air display held for Peruvian independence celebrations, traditionally held on July 29 of each year. (IEHAP)

Although of poor quality this picture shows a SE.3160 equipped with napalm containers during exercises held on the Reque gunnery range in Chiclayo during the early seventies. (IEHAP)

On December 18, 1973, Peruvian EP Gen. Augusto Freyre Garcia was flying Alouette III number FAP 6169 belonging to EA 332 on route to the San Juan del Oro province, in Puno, in order to review the works on a highway at that location being performed by EP personnel. However, 20 minutes after leaving Arequipa the pilot reported a mechanical failure and attempted an emergency landing at "Pampa de Sumbay" field but, just as the helicopter was approaching the ground, the engine lost power and the aircraft plunged to earth, the rotors hitting the ground and breaking, piercing the crew compartment and instantly killing Freyre, while severely injuring the pilot, *Alf.* FAP Gonzalo Gaviola Tejada and his mechanic, 1st Class S.O. Elías Cano Paredes, who died from their wounds shortly after despite the first aid provided by the local population.

Another SE.3160 was lost on March 15, 1983, when FAP 640[9] crashed in Cusco due an engine failure while carrying a television crew. No causalities were recorded but the aircraft was damaged beyond repair.

Despite these unfortunate incidents the SE.3160 continued to soldier with EA 332 until mid nineties when lack of spare parts forced the progressive grounding of the remaining aircraft. With the arrival of more capable aircraft the Air Force sold its remaining Alouettes and by the early nineties at least three of the five airworthy survivors had bend sold to civilian operators. The last airworthy FAP Alouette III was finally grounded in 1998 and later scrapped.

AEROSPATIALE AS.350B-1 ECUREUIL

In May 1989 Peruvian air force purchased a single Aerospatiale AS.350B-1 for evaluation purposes. The single Ecureuil[10] was delivered by the end of the year and received FAP 661 before being assigned to the EA 332 where it served as a VIP transport. The AS.350 didn't live up to FAP expectations and it was withdrawn from active service during late 1993 and, later, sold to a French civilian customer in the year 1995. The aircraft currently operates under the civilian registry F-GJAQ.

9 Co/Nu 2018.

10 Co/Nu 2177, French civil registry F-WZFP and Peruvian civil registry OB-1680.

A rare in flight view of the single AS.350B-1 Ecureuil helicopter operated by the FAP. The aircraft, identified with the number FAP 661, didn't live up to the expectations, and was sold in the early nineties.

DASSAULT FALCON 20F

A single Dassault Falcon 20F VIP aircraft was purchased in 1982 and delivered during 1983. This aircraft, identified with construction number 434 and French registry F-WRQP, was assigned with number FAP 300 and Peruvian civil registry OB-1433. The aircraft was immediately assigned to the EA 335 where it served both as a VIP transport as well as in photo reconnaissance duties.

This aircraft's operational life came to an end on June 18, 2010, when the aircraft suffered an engine failure shortly after takeoff from "Capitan Jose Abelardo Quiñones" AB in Chiclayo, forcing its pilot, Cmdr. Nelson Soberón Salas, to perform an emergency landing just outside the base, luckily with no human losses. The aircraft, however, was damaged beyond repair and had to be withdrawn from service.

Delivered in 1983 as part of the Jupiter contracts, the Falcon 20F was a capable VIP and fast courier aircraft. It was assigned to the Servicio Aerofotográfico Nacional. (IEHAP)

A nice view of the Dassault Aviation Falcon 20F taken at Grupo Aéreo N° 8 in Callao in October 2008. The aircraft received a new paint job during the repairs needed by the damages caused by a minor accident in 2004 and was written off wearing this livery in Chiclayo on June 18, 2010. (DIRIN)

CHAPTER FIVE
TECHNICAL CHARACTERISTICS

CAUDRON G.3

Length:	6.40 m
Span:	13.40 m
Height:	2.50 m
Wing span:	27.00 m²
Empty weight:	420 kg
Max weight:	734 kg
Power plant:	Le Rhone C 80 hp
Max Speed:	112 km/h
Ceiling:	4,300 m
Endurance:	four hours.

SALMSON 2A.2

Length:	8.53 m
Span:	11.76 m
Height:	2.9 m
Wing area:	37.27 m²
Empty weight:	780 kg
Max weight:	1,290 kg
Power plant:	Salmson 9z 260 hp radial engine.
Max speed:	188 km/h at sea level
Ceiling:	6,250 m
Range:	500 km (Two auxiliary fuel tanks could be installed in supports located between the wings to increase range to 700 km)
Armament:	One forward firing Vickers 0.303 in machinegun Two rear firing Lewis 0.303 in machineguns mounted on a Scarf ring.

HANRIOT H.438

Length:	8.12 m
Height:	3.29 m
Wing span:	11.75 m
Wing area:	30.24 m²
Empty weight:	1,032 kg
Max weight:	1,410 kg
Power plant:	Salmson 9c 330 hp radial engine.
Max speed:	186 km/h at sea level
Armament:	One rear firing Lewis 0.303 in machinegun mounted on a Scarf ring.

HENRI FARMAN III

Length:	13.45 m
Height:	3.48 m
Wing span:	10.50 m
Wing area:	42.00 m²
Empty weight:	320 kg
Max weight:	550 kg
Power plant:	50 hp Vivinus engine.
Max speed:	60 km/h at sea level

DASSAULT-MIRAGE 5P

Length:	15.56 m
Height:	4.5 m
Wing span:	8.2 m
Wing area:	35 m²
Empty weight:	7,150 kg
Max weight:	13,700 kg
Power plant:	SNECMA Atar 09-C turbojet
Dry trust:	4,278 KG
Trust with afterburner:	6,200 kgt
Max speed:	Mach 2.2 at 12.000 m
Service Ceiling:	18.000 m

DASSAULT-MIRAGE 2000P

Length: 14.36 m
Height: 5.20 m
Wing span: 9.13 m
Wing area: 41 m^2
Empty weight: 7,500 kg
Max weight: 13,800 kg
Power plant: SNECMA M53P-2 turbofan
Dry trust: 6,577 KG
Trust with afterburner: 9,707 KG
Max speed: Mach 2.2 at 12.000 m
Service Ceiling: 17.069 m

MORANE-SAULNIER MS.231ET.2

Length: 7.51 m
Height: 2.63 m
Wing span: 12.00 m
Wing area: 37.27 m^2
Empty weight: 829 kg
Max weight: 1,150 kg
Power plant: Lorraine Mizar 7 Mb 240 hp radial engine
Max speed: 205 km/h at sea level
Service Ceiling: 5000 m

MORANE-SAULNIER MS.316ET.2

Length: 7.51 m
Height: 2.63 m
Wing span: 12.00 m
Wing area: 21.64 m^2
Empty weight: 672 kg
Max weight: 933 kg
Power plant: Lorraine 5 Pa 120 hp 5 cylinder radial engine (later replaced by a 125 hp Kinner B-5 5 cylinder radial engine)
Max speed: 130 km/h at sea level
Service Ceiling: 3,500 m

VOISIN TYPE BORDEAUX

Length: 10.00 m
Height: 2.63 m
Wing span: 11.00 m
Wing area: 37.27 m^2
Empty weight: 400 kg
Max weight: 680 kg
Power plant: 50 hp Gnome 7cylinder radial engine
Max speed: 188 km/h at sea level

POTEZ 390/391bis

Span: 16 m
Length: 9.90 m
Height: 3.40 m
Surface: 35 m^2.
Power plant: (Potez 390) 550 hp Lorraine-Dietrich 12Hars; (Potez 391A.2Bis) 720 hp Lorraine-Dietrich 12Hdrs
Maximum speed: 295 km/h at 3500 m
Cruise speed: 240 km/h at the same altitude.
Service ceiling: 9000 m
Range: 760 km
Armament. Two Darne 7.65 mm machineguns fitted on a ring mount; one fixed forward firing Darne 7.65 mm machine gun firing through the propeller. Up to 120 kg of bombs could be fitted either on special racks located under the wings or inside a special bomb bay located in the front fuselage.

NIEUPORT-DELAGE NiD.121C.1

Power plant: (NiD.121C.1) 550 hp Lorraine-Dietrich 12Hars; (NiD 123) 720 hp Lorraine-Dietrich 12Hdrs
Characteristics (121C.1)
Maximum speed: 368 km/h
Minimum speed: 115 km/h
Ceiling: 9900 m
Range: 400 km.
Height: 3.025 m
Wing span: 13 m
Length: 7.18 m.
Wing Surface: 22 m^2

Empty weight: 1433 kg
Maximum weight: 1866 kg.
Maximum at take off: 2000 kg.
Armament: four Vickers 7.65 mm machine guns (two synchronized shooting through the propeller arch, plus two others in the wings). Up to 120 kg of bombs could be carried in racks located under the wings.

GEORGES-LEVY GL.40HB.2

Length: 12.40 m
Height: 3.63 m
Wing span: 18.50 m
Wing area: 37.27 m²
Empty weight: 2050 kg
Max weight: 2350 kg
Crew: 3
Power plant: 280 hp Renault 12Fe in-line engine
Max speed: 150 km/h

SUD SE.3130 "ALOUETTE II"

Crew: 1
Capacity: 4
Power plant (on Peruvian machines): 550 shp (410 kW) Turbomeca Astazou IIA
Maximum speed: 185 km/h
Cruise speed: 170 km/h
Ceiling: 2300 m
Range: 565 km
Endurance: 4.1 hours
Height: 2.75 m
Rotor diameter: 10.20 m
Length: 9.66 m
Disc area: 81.7 m²
Empty weight: 895 kg
Maximum at take off: 1600 kg.

SUD SE.3160 "ALOUETTE III"

Crew: 2
Capacity: 5
Power plant (on Peruvian machines): 870 shp (649 kW) Turbomeca Astazou XIV turbo shaft engine, derated to 447 kW (660 shp).
Maximum speed: 210 km/h
Cruise speed: 185 km/h
Ceiling: 3200 m
Range: 540 km
Height: 3.00 m
Rotor diameter: 11.02 m
Length: 10.03 m
Disc area: 95.38 m²
Empty weight: 1143 kg
Maximum at take off: 2200 kg.
Crew: 1
Capacity: 4